Applied Dermatotoxicology:
Clinical Aspects

Applied Dermatotoxicology: Clinical Aspects

Howard Maibach
Department of Dermatology, California Medical School,
San Francisco, California, USA

Golara Honari
Department of Dermatology, University of California San Francisco,
San Francisco, California, USA

and

Department of Dermatology, Cleveland Clinic, Cleveland, Ohio, USA

AMSTERDAM • BOSTON • HEIDELBERG • LONDON
NEW YORK • OXFORD • PARIS • SAN DIEGO
SAN FRANCISCO • SINGAPORE • SYDNEY • TOKYO

Academic Press is an imprint of Elsevier

Academic Press is an imprint of Elsevier
225 Wyman Street, Waltham, MA 02451, USA
525 B Street, Suite 1800, San Diego, CA 92101, USA
The Boulevard, Langford Lane, Kidlington, Oxford, OX5 1GB, UK
32 Jamestown Road, London NW1 7BY, UK
Radarweg 29, 1043 NX Amsterdam, The Netherlands

British Library Cataloguing in Publication Data
A catalogue record for this book is available from the British Library

Library of Congress Cataloging-in-Publication Data
A catalog record for this book is available from the Library of Congress

ISBN: 978-0-12-420130-9

For information on all Academic Press publications
visit our website at **store.elsevier.com**

This book has been manufactured using Print On Demand technology.

ELSEVIER · Book Aid International · **Working together to grow libraries in developing countries**

www.elsevier.com • www.bookaid.org

CONTENTS

Angelova-Fischer, Irena, MD, PhD
Department of Dermatology, University of Lübeck, Germany

Honari, Golara
Department of Dermatology, University of California San Francisco,
San Francisco, California, USA; Department of Dermatology, Cleveland Clinic,
Cleveland, Ohio, USA

Ibbotson, Sally Helen
Photobiology Unit, Dermatology Department, University of Dundee and Ninewells
Hospital & Medical School, Dundee, Tayside, Scotland, United Kingdom

Irfan, Mahwish
Cleveland Clinic Department of Dermatology, Cleveland, Ohio, USA

Lachapelle, Jean-Marie, MD, PhD
Catholic University of Louvain, Cliniues Universitaircs Saint-Luc, Avenue
Hippocrate, Brussels, Belgium

Maibach, Howard
Department of Dermatology, California Medical School, San Francisco,
California, USA

Mateeva, Valeria, MD
Department of Dermatology, Military Medical Academy, Sofia, Bulgaria

This book offers an introduction to disciplines in dermatotoxicology; the science that studies the effect of environmental physical or chemical elements affecting skin along with their clinical consequences. For detailed information, we refer the readers to the textbook of *Dermatotoxicology*, 8th edition.

We provide concise information on clinical presentations and morphologic characteristics of distinct clinical entities, caused by environmental exposures, as well as latest toxicologic methods. We have tried to present the science of dermatotoxicology in a practical and easy to follow format, aiming to discuss clinical consequences of exposure to offending agents along with the main predictive methods for the assessment of these agents.

We thank all authors and our publisher with special thanks to Ms. Shannon Stanton and Ms. Priya Kumaraguruparan from Elsevier for their immense dedication.

We hope you find this book a useful introduction and appreciate your comments and suggestions.

Golara Honari
Howard I. Maibach

CHAPTER 1

Skin Structure and Function

Golara Honari and Howard Maibach

INTRODUCTION

Skin as one of the largest organs in body has multiple key features required to interact dynamically with the environment. Primary functions include a barrier function against environmental hazards such as ultraviolet (UV) radiation, chemical and physical insults, and microorganisms. Skin also prevents dehydration, regulates temperature, and has self-healing properties. Dynamic and complex arrangement of variety of cells, element of the extracellular matrix, vascular, appendageal, and nervous structures each play a role. Knowledge of skin penetration pathways is essential in assessment of chemical safety, drug delivery systems, and formulation of cosmetic products.

These chapter overviews essential structural elements of skin, their key functions relevant to dermatotoxicology, skin absorption pathways, and key devices and methods to measure skin properties.

STRUCTURE AND FUNCTION

Normal skin consists of three main layers: epidermis, dermis, and hypodermis.

*   *Epidermis*: Barrier function, innate immunity, UV protection.
*   *Dermis*: The largest component of skin, dermis is an integrated system of fibrous cellular and acellular matrix. Many cell types reside in dermis including fibroblasts, macrophages, mast cells. Vascular, lymphatic, and nervous networks are present in dermis.
*   *Hypodermis (subcutis)*: Provides mechanical and physiologic support and contains larger source of vessels and nerves.

There are regional variations in the skin thickness and presence of different appendages such as hair, sebaceous, and sweat glands, which can

Applied Dermatotoxicology. DOI: http://dx.doi.org/10.1016/B978-0-12-420130-9.00001-3

affect the functional properties. For example, hair baring skin is typically thinner and more permeable than nonhair baring skin of palms and soles.

Epidermis

Epidermis is the outermost layer and is about 0.05−1 mm in thickness depending on body part. Three main populations of cells reside in the epidermis: keratinocytes, melanocytes, and Langerhans cells. Keratinocytes are the predominant cells in the epidermis, which are constantly generated in the basal lamina and go through maturation, differentiation, and migration to the surface. As keratinocytes differentiate, they form three layers above the basal layer known as stratum spinosum (SP), stratum granulosum (SG), and stratum corneum (SC) (Figure 1.1). Keratinocyte transit time from basal layer up to SC is about 14 days[1] and turn over time within SC is also around 14 days,[2] certain inflammatory conditions can affect these turn over times.

SC is the outer layer of the epidermis and serves as the main functional barrier. A theoretical model is "brick and mortar" like structure where bricks represent terminally differentiated nonviable keratinocytes, also known as corneocytes embedded in intercellular lipid membranes.[3] As corneodesmosomes (protein bridges between corneocytes) degrade, lacunar spaces are created within the SC referred to as

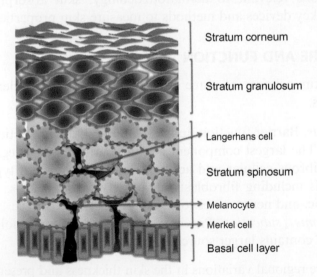

Stratum corneum

Stratum granulosum

Langerhans cell

Stratum spinosum

Melanocyte

Merkel cell

Basal cell layer

Figure 1.1 Schematic of epidermis—basal cell layer is the deepest layer of epidermis differentiating to spinous cells then to granular cells and eventually terminally differentiate to SC.

aqueous "pore" pathway. These spaces can extend and form continues networks, creating a pathway for penetration across the SC.[4]

Major components of the SC lipid membranes are free fatty acids, ceramides, and esterols.[5] Melanocytes are neural crest-derived, pigment synthesizing dendritic cells that reside primarily in the basal layer. Merkel cells are mechanosensory receptors also present in basal layer. Langerhans cells are dendritic antigen-processing and antigen-presenting cells in the epidermis.[6] They form 2–8% of the total epidermal cell population, mostly found in a suprabasal position. The dermal–epidermal junction (DEJ) is a basement membrane zone that forms the interface between the epidermis and dermis. The major functions of the DEJ are to attach the epidermis and dermis to each other and to provide resistance against external shearing forces.

Dermis

The dermis is an integrated system of fibrous cellular and acellular matrix that accommodates nervous and vascular structures as well as epidermally derived appendages. Many cell types reside in the dermis including fibroblasts, macrophages, mast cells, and circulating immune cells. Dermis is responsible for skin elasticity, pliability, and tensile strength and provides protection against mechanical injury, retins water, and aids in thermal regulation. Dermis also contains and supports receptors of sensory stimuli and a key element in wound healing.[7]

Hypodermis

Hypodermis is primarily composed of adipose tissue, which insulates the body, serves as a reserve energy supply. It cushions and protects skin and supports nerves, vessels, and lymphatics located within the septa, supplying the overlying region.

SKIN APPENDAGES

Skin appendages include nails, hair, sebaceous glands, eccrine (sweat) glands, and apocrine glands. They have two distinct components: superficial and deeper components in the dermis, which are down growths of epidermis. Dermal component regulates differentiation of the appendage. During embryonic development, dermal–epidermal interactions are critical for the induction and differentiation of these structures.

SKIN A ROUTE OF ENTRY

The SC is 3–20 μm in thickness, composed of 15–25 layers of corneocytes. It provides an effective barrier against transcutaneous water loss and entry of exogenous materials.

Extracellular lipids contribute to barrier function and the route taken through the SC by all molecules. Arrangement of extracellular lipids, their hydrophobicity, composition, and distribution of key components (ceramides, cholesterol, and free fatty acids) provide more barrier function.[8] Skin absorption varies between different body parts, and between individuals, these regional intra- and inter-individual variations are partly related to variations in lipid composition and SC thickness.[8] A range of biological factors can influence the rate and extent of percutaneous penetration including anatomical site, age, appendageal density, SC morphology, and composition. Routs through which chemicals can cross the SC (Figure 1.2) include[9]:

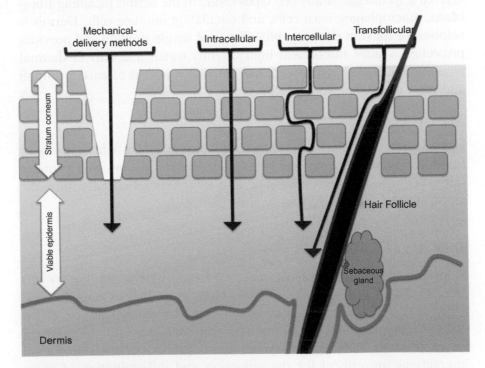

Figure 1.2 Schematic pathways of penetration into the skin with arrangement of corneocytes in a "brick and mortar" model.[9]

- Intercellular (or extracellular), in which chemicals pass exclusively through the lipid matrix
- Intracellular or transcellular, in which chemicals pass through both the lipid matrix and the corneocytes themselves.
- Through skin appendages
- Mechanical methods to remove SC such as stripping, ablation, micro needles, etc.

In vivo measurements skin absorption to quantify and rate the extent of absorption is of fundamental importance in the risk assessment of compounds that are active via the dermal route of entry, though the details are beyond the scope of this chapter.

MEASURING SKIN

Measuring physical characteristics of skin using biophysical instruments provides key information about various skin parameters. Many noninvasive techniques and equipment are available with increasing applications within dermatotoxicology, allowing study of skin in real time and providing objective, quantitative data.

Parameters measured with these techniques provide information about a particular aspect of skin. Using multiple parameters measured simultaneously, along with clinical assessment provides more comprehensive analysis. Histologic studies may be used to complement these analyses. Few parameters and techniques are briefly introduced in this chapter; additional texts are available for comprehensive study.[10,11]

Skin Surface pH

Skin pH is normally acidic, ranging in pH values of 4−6, while the PH in body's internal environment is near-neutral, ranging from 7 to 9.[12] The term "acid mantle" refers to inherent acidic nature of the SC. Skin pH affects barrier function and SC cohesion. Elevation of pH in normal skin creates a disturbed barrier.[13] Measurement of skin surface pH is used to assess the acidity of the skin's surface which skin surface pH can vary according to the time of day, skin site, and between individuals. There are several instruments available for measurement of skin pH; basically any standard, portable pH meter with a planar electrode should suffice.

Sebum

Sebum is a light yellow viscous fluid, composed of triglycerides, free fatty acids, squalene, wax and sterol esters, and free sterols. Sebum is produced by sebaceous glands and contributes to moisture balance in the SC. Sebum production is mainly influenced by androgens and varies among individuals and races, but the average rate in adults is approximately $1 \, mg/10 \, cm^2$ every 3 h.[14] Sebum production less than $0.5 \, mg/10 \, cm^2$ every 3 h is associated with dry skin, and values of $1.5-4.0 \, mg/10 \, cm^2$ every 3 h associated with seborrhea.[15] Sebum can be measured using a gravimetric or a variety of photometric techniques.

Methods used in the past for measurement of skin lipids included swabbing skin by absorbing pads followed by soaking them in solvents or by weighing the absorptive tapes (gravimetric analysis).[16] Also photometric analyses such as "grease-spot" photometry or UV–Visible spectrophotometry are used. Photometric techniques can provide additional information, such as droplet size and distribution.[8] Multiple commercially available devices can provide quantitative measures of sebum secretion.

Desquamation

Loss of superficial SC is a natural process called desquamation, which can be affected in certain disease. Desquamation is used to assess structure and biological dynamics of SC and to investigate effect of drugs and topical products on skin. Techniques for desquamation measurement usually involve stripping surface layers for analysis or visual assessment through skin imaging.

Squamometry involves striping of corneocytes form surface of SC using adhesive disks followed by staining with toluidine blue-basic fuchsin ethanol-based solution and reading the colorimetric variable.[17] Image analysis systems are also used to asses desquamation.[18–20] These techniques are rapid, reliable, and sensitive.

Thickness

Epidermal and SC thickness can be measured in vivo, using noninvasive methods such as ultrasound-based devices, confocal microscopy and spectroscopic techniques.[21–25] These measurements have investigative and clinical applications.

The Measurement of Transepidermal Water Loss

Transepidermal water loss (TEWL) is the amount of water that passively evaporates through skin to the external environment due to water vapor pressure gradient on both sides of the skin barrier and is used to characterize skin barrier function. The average TEWL in human is about 300−400 mL/day; however, it can be affected by environmental and intrinsic factors. In high humidity, the amount of water loss will decrease due to the drop in the water vapor pressure gradient. TEWL varies in different anatomic sites and is inversely related to the corneocyte size. Skin sites with smaller corneocytes have higher TEWL values.[26−28] Multiple instruments are commercially available to measure TEWL, providing valuable data with applications in clinical settings, toxicology, and product development. TEWL is a sensitive indicator of skin irritation and is widely used in objective analysis of irritancy potential or protective properties of topical products. The accuracy of TEWL measurements can be influenced by environmental factors such as humidity, temperature, ventilation, and intrinsic factors. It is essential that these measurements be conducted under standard conditions.[29]

Hydration Measurement

Water content of SC in normal conditions is estimated around 5−10% in the outer layer and about 30% near viable epidermis.[30] Water contents of skin influence its physical properties, viscoelastic characteristics, and functional properties such as the drug penetration and barrier function. In disease states such as atopic dermatitis, SC cannot stay properly hydrated. Hydration status of SC is considered a measure of toxicity. Hydration measurements are extensively used in assessment of topical product safety and efficacy.

Various methods have been introduced to indirectly measure hydration. One method uses electrical properties of skin like capacitance, impedance, resistance, and conductance, to calculate hydration levels.[31] Another method measures transient thermal transfer (TTT), by using a probe to transfer a constant generated thermal pulse to the epidermis, while simultaneously obtains high precision measurements of skin temperature. Water content of SC is calculated based on changes detected in the temperature.[32] The third technique, nuclear magnetic resonance (NMR) spectroscopy, directly measures the total water content of the epidermis and the outer dermis, using magnetic fields.[33] NMR provides direct measurements and is considered a reference technique for skin hydration studies. However, all the available techniques can complement each other.[34]

Measurement of Vascular Perfusion

Skin microcirculation and perfusion can be affected by a number of exogenous and endogenous factors. Changes in cutaneous vascular perfusion may reflect a physiologic response such as thermoregulation, or a pathologic response such as inflammation caused by exposure to chemical irritants and concomitant release of inflammatory mediators. Also exposure to topical vasoactive drugs can affect skin circulation. Laser Doppler instruments measure the Doppler shift induced by the laser light that's being scattered by moving red blood cells. A signal is quantified based on average red blood cell concentration and velocity. Since this measure is not an exact measure, it is referred to as flux, which has a linear relationship with the actual flow.[35–37] Two distinct laser Doppler tools are used: laser Doppler flowmetry (LDF), which has a small probe touching the skin, measuring blood flow over a small volume (1 mm^3 or smaller). The second method is laser Doppler imaging (LDI), in which the laser beam is emitted at a certain distance above the skin surface and reflected by a computer-driven mirror to scan an area of skin. DLI provides two-dimensional images mapping the perfusion but is a slow method. Quantifying fast changes in blood flow is easier with DLF compared to DLI.[35]

Laser speckle contrast imaging (LSCI) is a more recent, noncontact imaging technique that provides information on skin perfusion over large areas (up to 100 cm^2) with a high frequency (up to 100 images/s). These images are obtained based on the generation of a high contrast grainy pattern when laser hits a matt surface. This pattern is called speckled pattern, which fluctuates when objects move. Images created by LSCI reflect fast changes in skin blood flux over wide skin areas.[35,38]

These techniques along with other methods briefly mentioned in this chapter are among many tools to assess skin parameters and to provide objective measures for investigators and clinicians. Many bioengineering methods are subject to operational guidelines and have limitations. Data obtained from various methods can complement each other.

REFERENCES

1. Weinstein G, Van Scott E. Autoradiographic analysis of turnover times of normal and psoriatic epidermis. *J Invest Dermatol.* 1965;45(4):257–262.

2. Bergstresser P, Taylor J. Epidermal "turnover time"—a new examination. *Br J Dermatol.* 1977;96:503–509.

3. Michaels AS, Chandrasekaran SK, Shaw JE. Drug permeation through human skin. Theory and in vitro experimental measurements. *AICHE J.* 1975;21(5):985–996.

4. Menon GK, Elias PM. Morphologic basis for a pore-pathway in mammalian stratum corneum. *Skin Pharmacol.* 1997;10(5–6):235–246.

5. Lee D, Ashcraft JN, Verploegen E, Pashkovski E, Weitz DA. Permeability of model stratum corneum lipid membrane measured using quartz crystal microbalance. *Langmuir.* 2009; 25(10):5762–5766.

6. Mutyambizi K, Berger CL, Edelson RL. The balance between immunity and tolerance: the role of Langerhans cells. *Cell Mol Life Sci.* 2009;66(5):831–840.

7. Chu DH. *Development and structure of skin. Fitzpatrick's Dermatology in General Medicine.* 8th ed., USA: McGraw-Hill; 2012.

8. Elias PM. The epidermal permeability barrier: from the early days at Harvard to emerging concepts. *J Invest Dermatol.* 2004;122(2):xxxvi–xxxix.

9. Prausnitz MR, Mitragotri S, Langer R. Current status and future potential of transdermal drug delivery. *Nat Rev Drug Discov.* 2004;3(2):115–124.

10. Fluhr WJ, Elsner P, Berardesca E, Maibach HI. *Bioengineering of the skin: water and the stratum corneum. Dermatology: Clinical & Basic Science.* 2nd ed. USA: CRC Press; 2004.

11. Agache P, Humbert P. *Measuring the Skin.* Heidelberg: Springer; 2004.

12. Ali SM, Yosipovitch G. Skin pH: from basic science to basic skin care. *Acta Derm Venereol.* 2013;93(3):261–267.

13. Hachem JP, Crumrine D, Fluhr J, Brown BE, Feingold KR, Elias PM. pH directly regulates epidermal permeability barrier homeostasis, and stratum corneum integrity/cohesion. *J Invest Dermatol.* 2003;121(2):345–353.

14. Plewig G, Kligman A. *Acne and Rosacea.* Berlin: Springer; 2000.

15. Bolognia JL, Jorizzo JL, Schaffer JV. *Dermatology.* 3rd ed. China: Elsevier; 2012.

16. Pande SY, Misri R. Sebumeter. *Indian J Dermatol Venereol Leprol.* 2005;71(6):444–446.

17. Pierard-Franchimont C, Henry F, Pierard GE. The SACD method and the XLRS squamometry tests revisited. *Int J Cosmet Sci.* 2000;22(6):437–446.

18. Black D, Boyer J, Lagarde JM. Image analysis of skin scaling using D-squame samplers: comparison with clinical scoring and use for assessing moisturizer efficacy. *Int J Cosmet Sci.* 2006;28(1):35–44.

19. Wilhelm KP, Kaspar K, Schumann F, Articus K. Development and validation of a semiautomatic image analysis system for measuring skin desquamation with D-squames. *Skin Res Technol.* 2002;8(2):98–105.

20. Xhauflaire-Uhoda E, Loussouarn G, Haubrechts C, Leger DS, Pierard GE. Skin capacitance imaging and corneosurfametry. A comparative assessment of the impact of surfactants on stratum corneum. *Contact Dermatitis.* 2006;54(5):249–253.

21. El Gammal S, El Gammal C, Kaspar K, et al. Sonography of the skin at 100 MHz enables in vivo visualization of stratum corneum and viable epidermis in palmar skin and psoriatic plaques. *J Invest Dermatol.* 1999;113(5):821–829.

22. Kaspar K, Vogt M, Ermert H, Altmeyer P, el Gammal S. 100 MHz sonography in the visualization of the palmar stratum corneum after application of various creams and ointments. *Ultraschall Med.* 1999;20(3):110−114.

23. Corcuff P, Bertrand C, Leveque JL. Morphometry of human epidermis in vivo by real-time confocal microscopy. *Arch Dermatol Res.* 1993;285(8):475−481.

24. Caspers PJ, Lucassen GW, Carter EA, Bruining HA, Puppels GJ. In vivo confocal Raman microspectroscopy of the skin: noninvasive determination of molecular concentration profiles. *J Invest Dermatol.* 2001;116(3):434−442.

25. Egawa M, Hirao T, Takahashi M. In vivo estimation of stratum corneum thickness from water concentration profiles obtained with Raman spectroscopy. *Acta Derm Venereol.* 2007;87(1):4−8.

26. Rougier A, Lotte C, Corcuff P, Maibach HI. Relationship between skin permeability and corneocyte size according to anatomic site, age and sex in man. *J Soc Cosmet Chem.* 1988;39:15−26.

27. Machado M, Salgado TM, Hadgraft J, Lane ME. The relationship between transepidermal water loss and skin permeability. *Int J Pharm.* 2010;384(1−2):73−77.

28. Hadgraft J, Lane ME. Transepidermal water loss and skin site: a hypothesis. *Int J Pharm.* 2009;373(1−2):1−3.

29. Sotoodian B, Maibach HI. Noninvasive test methods for epidermal barrier function. *Clin Dermatol.* 2012;30(3):301−310.

30. Blank IH, Moloney III J, Emslie AG, Simon I, Apt C. The diffusion of water across the stratum corneum as a function of its water content. *J Invest Dermatol.* 1984;82(2):188−194.

31. Berardesca E, Borroni G. Instrumental evaluation of cutaneous hydration. *Clin Dermatol.* 1995;13(4):323−327.

32. Berardesca E, Fideli D, Borroni G, Rabbiosi G, Maibach H. In vivo hydration and water-retention capacity of stratum corneum in clinically uninvolved skin in atopic and psoriatic patients. *Acta Derm Venereol.* 1990;70(5):400−404.

33. Foreman MI. A proton magnetic resonance study of water in human stratum corneum. *Biochim Biophys Acta.* 1976;437(2):599−603.

34. Girard P, Beraud A, Sirvent A. Study of three complementary techniques for measuring cutaneous hydration in vivo in human subjects: NMR spectroscopy, transient thermal transfer and corneometry—application to xerotic skin and cosmetics. *Skin Res Technol.* 2000; 6(4):205−213.

35. Roustit M, Cracowski JL. Assessment of endothelial and neurovascular function in human skin microcirculation. *Trends Pharmacol Sci.* 2013;34(7):373−384.

36. Stern MD. In vivo evaluation of microcirculation by coherent light scattering. *Nature.* 1975;254(5495):56−58.

37. Ahn H, Johansson K, Lundgren O, Nilsson GE. In vivo evaluation of signal processors for laser Doppler tissue flowmeters. *Med Biol Eng Comput.* 1987;25(2):207−211.

38. Briers JD. Laser Doppler, speckle and related techniques for blood perfusion mapping and imaging. *Physiol Meas.* 2001;22(4):R35−R66.

CHAPTER 2

Irritant Contact Dermatitis: Clinical Aspects

Valeria Mateeva, MD and Irena Angelova-Fischer, MD, PhD

INTRODUCTION

Irritant contact dermatitis (ICD) is a common inflammatory skin condition in result of single or repetitive exposure to chemicals with irritant properties. Irritants interact with different components of the skin and elicit a broad spectrum of reactions that may range from mere sensory responses to a widespread severe disease with systemic involvement. The present chapter provides an overview of the clinical manifestations of ICD based on the course of disease and morphology.

CLINICAL MANIFESTATIONS OF IRRITANT DERMATITIS

Irritant Reaction

Prolonged exposure to mild irritants such as water, soaps, detergents, and solvents may induce a monomorphic skin response described as irritant reaction. Wet work in at risk occupations such as hairdressers, food processing industry, and metal workers is a significant factor for development of irritant reactions, in particular in the early stage of occupational training.[1-3] The initial clinical manifestations may be discrete and include erythema, scaling, vesiculation, or erosions affecting the webs of the fingers and dorsum of the hands. The irritant reaction may progress to manifest chronic irritant dermatitis or resolve spontaneously despite continuous exposure (hardening phenomenon).[4-6]

Acute Irritant Dermatitis

Acute ICD develops as a result of a single exposure to an irritant in sufficient concentration and for a sufficient time to elicit an inflammatory reaction.[7] The clinical signs appear minutes to hours after exposure and may vary from a mild irritant reaction limited to subtle erythema, skin

Applied Dermatotoxicology. DOI: http://dx.doi.org/10.1016/B978-0-12-420130-9.00002-5

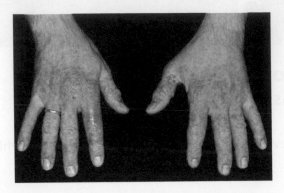

Figure 2.1 Acute irritant dermatitis: Erythema, edema, oozing, erosions, scaling, and yellowish crusts after exposure to undiluted surface disinfectant.

dryness, and chapping to a florid dermatitis with edema, vesiculation, and oozing/exudation (Figure 2.1). Immediate sensation of burning, stinging, itch, or pain are common, associated subjective symptoms. The skin lesions are usually asymmetric in distribution, well demarcated, and limited to the area of the contact with the irritant. The rapid onset of symptoms together with the pattern and distribution of the clinical manifestations in relationship to irritant exposure are key orientation points for correct diagnosis.

Cessation of the contact with the causative agent results in most cases in fast improvement of the clinical signs and symptoms consistent with a "decrescendo" pattern of reactivity. As long as the damage caused by the irritant is limited to the epidermis and reversible, acute ICD has a generally good prognosis and tends to resolve without sequellae within days or weeks.[8,9] The persistence of symptoms after elimination of the contact with the causative agent, however, may necessitate performance of further diagnostic tests to differentiate irritant from allergic contact dermatitis (ACD). In contrast to ACD, irritant dermatitis does not require prior sensitization and the patch tests remain negative.

Delayed Acute Irritant Dermatitis
Certain types of irritants such as hexanediol diacrylate, butanediol diacrylate, dithranol, and benzalkonium chloride, among others, may elicit a delayed skin response with onset of the inflammatory reaction up to 24 h after exposure (delayed acute irritant dermatitis).[5,10,11] The manifestations of the delayed reaction are similar to acute ICD however follow a "crescendo" pattern of reactivity characterized by a transient increase of the clinical signs and symptoms over time. The unspecific and often

overlapping clinical findings as well as unusual course of the skin irritant reaction may thus mimic ACD and indicate the performance of a patch test to differentiate one condition from the other.[2,12–14] The constellation of a negative patch test result and exposure to a substance known to cause delayed irritancy support the diagnosis of delayed acute ICD. Though persistent for several days, delayed acute ICD has generally good prognosis.[10]

Chronic Cumulative ICD

Chronic cumulative ICD develops as a result of multiple subthreshold insults to the skin by weak irritant stimuli that cause perpetuous cell damage with subsequent release of proinflammatory mediators and gradual deterioration of the epidermal barrier function.[7,15] It is the most common type of irritant dermatitis and predominantly affects the hands, face, or the eyelids. Hand eczema is the most frequent form of disease with estimated point prevalence in the general population of 4.2%, 1-year prevalence of 9.7%, and lifetime prevalence of 15%, as shown by a detailed review of the literature published between 1964 and 2007.[16] In addition, chronic cumulative irritant dermatitis is the most common form of occupational contact dermatitis that accounts for 50–80% of the cases.[17–19]

The individual susceptibility for development of manifest chronic cumulative ICD shows considerable variations that may be attributed to genetic, host-related and environmental factors. Genetic markers for increased susceptibility to irritant dermatitis so far relate primarily to variations in the genes involved in inflammation and skin barrier function.[20,21] Transition polymorphisms at position −308 within the promoter region of the gene encoding the proinflammatory cytokine tumor necrosis factor-alpha (TNF-α) have been associated with enhanced reactivity to primary irritants and increased susceptibility to irritant dermatitis. Carriers of the rarer A allele, *TNF-308 A* (TNF2 allele) have increased production of TNF-α and in addition to a lower irritation threshold to model irritants such as sodium lauryl sulfate and benzalkonium chloride, have been shown to develop occupational irritant dermatitis at lower irritant exposure levels compared to carriers of *TNF-308 G* (TNF1 allele).[22–26] In contrast, variations in the gene encoding interleukin-1α (IL-1α) such as the presence of an *IL1A-889 T* allele have been shown to exert protective effects toward the development of manifest dermatitis that have been attributed to reduced IL-1α levels in the stratum corneum.[27,28]

Genetic variations related to the skin barrier that have been found to contribute to increased susceptibility to irritant dermatitis include loss-of-function mutations in the gene encoding the epidermal differentiation protein filaggrin (*FLG*). The *FLG* mutations are the most important individual risk factor for atopic dermatitis known so far,[29-31] whereas individuals with atopic skin disease have been found to show enhanced responses to irritant damage.[32-41] Individuals with atopic dermatitis and *FLG* mutations have an increased risk to acquire occupational ICD.[42-50] *FLG* mutation carrier state adjusted for atopic dermatitis has been found to increase the risk 1.6-fold, whereas the adjusted risk for atopic dermatitis was 2.9-fold. Furthermore, individuals with atopic dermatitis and *FLG* mutations have been identified to have an almost 5-fold risk to develop irritant dermatitis and defined, therefore as a highly susceptible group. In addition to increased susceptibility, *FLG* mutations carrier state has been shown to contribute to the persistence of hand eczema and confer an unfavorable disease outcome.[51,52]

Host-related factors, other than atopy, that have been studied in relationship and shown to influence the individual response to irritant damage and development of chronic irritant dermatitis include age, gender, skin type, psychological stress, anatomic site, and preexisting skin disease.[53-70] Though the properties of the irritant and the host-related factors are the major determinants for the outcomes of the irritant interaction with the skin, environmental variables such as temperature, air flow, ambient humidity, mechanical irritation, and occlusion may be significant yet frequently overlooked contributing factors to the development of chronic irritant dermatitis.[71-74]

Chronic irritant dermatitis often begins with a few circumscribed patches of skin dryness, scaling, mild erythema and chapping distally at the fingertips that subsequently spread to the webs of the fingers, the back of the hands, and later on to the palms. Characteristic clinical findings include erythema, skin dryness, scaling, hyperkeratosis, fissuring, and lichenification; vesiculation though sometimes present tends to be rather discrete and less typical (Figures 2.2 and 2.3). Occasionally, nummular lesions on the dorsa of the hands may be found. In contrast to acute ICD, the affected areas are less sharply demarcated and the clinical presentations are less polymorphic. Pain as a result of fissuring is a common subjective symptom and may considerably impair the daily activities as well as life quality.[75-77] Itching, burning, and skin tightness are the further commonly associated symptoms.

A considerable number of cases tend to be rather mild and the patients may easily overlook the link between exposure and clinical manifestations.

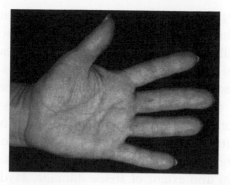

Figure 2.2 Cumulative occupational chronic irritant dermatitis in result of wet work and repeated exposure to food and food additives.

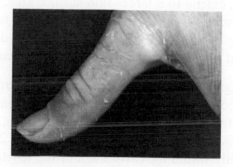

Figure 2.3 Chronic hand dermatitis after cumulative exposure to detergents and mechanical irritation.

The ubiquitous presence of irritants along with the long-standing subclinical inflammation, the overlap of clinical features, and frequent coexistence with atopic and/or ACD may delay and make the diagnosis of chronic ICD a challenge even to the experienced dermatologist. The persistence of clinical findings despite adequate treatment, skin protection, and skin care or the rapid worsening of symptoms necessitates further diagnostic steps to rule out skin conditions that may mimic, coexist, or superimpose on a preexisting disease.

Chemical Burns

Chemical burns are regarded as extreme variants of acute irritant dermatitis with irreversible tissue damage and necrosis.[78,79] Most chemical burns occur as a result of accidental exposure at the workplace. Nonwork-related cases are attributed mainly to exposure to corrosive agents in the household or during leisure activities.

Chemical burns are caused primarily by contact with strong acids or alkaline agents. Strong caustic agents cause progressive tissue damage that continues until there is no more unreacted chemical left or the agent has been neutralized by treatment. Different chemicals react with different intra- and intercellular components and consequently, the type and severity of tissue damage depend on the properties of the causative agent.[80] Most strong acids such as hydrochloric acid, hydrobromic acid, perchloric acid, nitric acid, and sulfuric acid cause protein denaturation and coagulation necrosis thus forming a barrier that limits the further spread of chemical and extent of tissue damage. Hydrofluoric acid, in contrast, exerts a different mode of action and causes the characteristic for alkalis liquefaction necrosis.[81] The liquefaction type of necrosis allows the further penetration of the chemical into the tissue that may continue for days, extend to a larger skin surface area, damage the deeper skin and musculoskeletal structures as well as result in systemic resorption and organ toxicity. Consistent with this mode of action, alkalis cause in general a more severe tissue damage compared to acids.

The clinical signs and symptoms of chemical burns appear within minutes after exposure and include painful erythema, burning, stinging, or smarting followed by development of blisters, erosions, ulcers, and necrotic areas surrounded by erythema (Figure 2.4). Wheals may be seen occasionally within the early stage of the reaction.[82] The causal relationship between exposure and onset of symptoms is usually clear; however, certain chemicals such as hydrofluoric acid or sulfur mustard

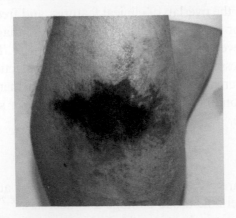

Figure 2.4 Chemical burn caused by accidental work-related exposure to sodium hydroxide. Photograph courtesy of Dr. T. W. Fischer, Department of Dermatology, University of Lübeck, Germany.

(2,2′-dichlorodiethyl sulfide) may give delayed reactions that begin hours up to a day after exposure.[79] Formation of eschars and the absence of blistering are characteristic for chemical burns induced by alkalis. In addition, chemical burns caused by alkalis or hydrofluoric acid are more painful than the ones caused by acidic agents.

Trapping of the toxicant by protective clothing and footwear prolongs exposure that may, along with occlusion, increase the capacity of weaker chemicals to induce tissue necrosis. Examples in that context are caustic burns in result of occupational exposure to wet cement.[83–94]

Dependent on the depth of tissue involvement chemical skin burns are classified into:

- **Superficial partial-thickness burns:** the damage extends to the level of the dermal papillae. The clinical manifestations include pain, erythema, edema, and occasionally blisters that resolve without scarring.
- **Deep partial-thickness burns:** the damage extends into the dermis and the lesions resolve with scarring.
- **Full-thickness burns:** the damage extends to the subcutaneous tissue. The lesions appear brownish-black or pale with an eschar. Function and/or sensation may be impaired and healing occurs with scarring.

Tissue damage limited to the epidermis may resolve without sequellae or result in postlesional hyper- or hypopigmentation. Chemical burns with involvement of the deeper skin layers heal ultimately with scar formation. Induction of contact sensitization is another important complication of chemical burns.[95–98]

SPECIAL CLINICAL FORMS OF IRRITANT DERMATITIS

Pustular and Acneiform Irritant Dermatitis

Pustular and acneiform dermatitis may develop after exposure to mineral oils and greases, coal tar, asphalt, chlorinated naphthalenes, polyhalogenated biphenyls, arsenic trioxide, metals, and metalworking fluids in occupational settings or as a result of consumer exposure to comedogenic constituents in cosmetic formulations.[99–103] Further etiologic factors include repeated or prolonged mechanical irritation through friction, rubbing, or pressure (acne mechanica).[99,104]

Oil acne is the most common form of occupational acneiform irritant dermatitis.[99] Occupations at particular risk include petroleum refiners, machine, automobile, and aircraft mechanics. Predilection sites for development of acneiform dermatitis are the areas in direct contact with the causative agent such as the dorsa of the hands, the forearms, and the face though indirect exposure through oil-soaked garments may produce lesions in protected skin areas.[105] The clinical manifestations of oil acne are polymorphic and include open and closed comedones, pustules, and inflamed nodules that may mimic severe forms of acne such as acne conglobata. Occupational exposure to known acnegens, unusual age of disease onset, and the involvement of skin areas beyond the predilection sites for acne suggest the correct diagnosis.

Chloracne develops as a result of occupational or environmental exposure to halogenated aromatic hydrocarbons with specific molecular configuration referred to as chloracnegens.[106,107] In different biological species chloracnegenic compounds target different organs however within the same species, the exerted toxic effects show similar patterns. Human skin is sensitive to chloracnegenic chemicals. Chloracne changes in skin may indicate a possible systemic toxicity.[108,109]

The primary lesions are multiple comedones affecting the malar crescents and the retroauricular area (Figure 2.5). In mild cases, these might be the only clinical manifestations, undistinguishable from common pathologies such as mild noninflammatory forms of acne or nodular elastosis with cysts and comedones (Favre–Racouchot syndrome).[99,106] The lesions become numerous with increasing severity and systemic toxicity and spread to the cheeks, forehead, neck, trunk,

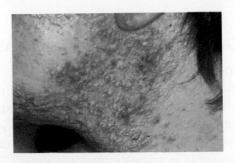

Figure 2.5 Chloracne: multiple comedones and cysts with characteristic involvement of the periauricular area and the neck after exposure to 2,3,7,8-tetrachlorodibenzo-p-dioxin (TCDD). Photograph courtesy of Prof. A. Geusau, Department of Dermatology, Medical University of Vienna, Austria.

and genital region, whereas the nose remains usually spared. The characteristic distribution of the skin lesions is of considerable diagnostic importance. The presence of pale, yellowish, or "straw-colored" cysts may further aid the diagnosis. Though inflammatory lesions may occur in progression, they are less characteristic, except for the most severe forms that mimic acne conglobata.

Further reported clinical manifestations of chloracne related to the skin and the skin appendages include facial erythema and edema, hyperpigmentation of the nails, lips, gingival, buccal and conjunctival mucosa, follicular hyperkeratosis, hypertrichosis, and palmoplantar keratoderma.[99,106,110]

Chloracne may have a prolonged clinical course and in severe cases the lesions may persist for years after cessation of exposure.[111-114]

Sensory Irritation

Sensory (subjective) irritation is defined as elicitation of an unpleasant sensation of burning, itching, stinging, smarting, or pain upon irritant contact with the skin in complete absence of clinical manifestations of dermatitis. The mechanisms underlying the enhanced sensory perception are incompletely understood[115-117] with newer studies focusing on the role of neurotrophins, endothelin receptors, and pain/cold/heat receptors of the transient receptor potential family (TRP) as well as polymorphisms at position -308 of TNF-α gene in the pathogenesis of sensitive skin.[25,118,119] The absence of visible signs upon clinical examination, the nonspecific character of symptoms, and variations in the individual's response to different stimuli make sensory irritation difficult to quantify and study. In 1977, Frosch and Kligman introduced the lactic acid stinging test as a model to study sensory irritation in a subset of individuals with sensitive skin referred to as "stingers."[120] The test relies on the application of lactic acid in the nasolabial fold and standardized grading of the intensity of symptoms compared to water used as control. Though the subjective symptoms induced by the application of lactic acid are reproducible, the test has limited value for predicting sensory irritation caused by unrelated chemicals.[121-123]

Airborne Irritant Dermatitis

Airborne irritant dermatitis is an inflammatory condition caused by skin contact with fibers, dust particles, and volatile compounds such as sprays, gasses, or vapors carried by the air.[124-129] The mode of exposure determines the characteristic pattern of distribution of the skin lesions with involvement of the face and neck as well as other uncovered body areas such as the hands and forearms. Contact with airborne irritants may occur in various occupational settings and most cases develop as a result of work-related exposure.[130] Trapping or accumulation of the causative agent under ill-fitting personal protective equipment may cause symptoms on covered areas and mask the link between exposure and disease manifestations. Airborne irritant dermatitis presents in most cases with erythema, discrete papules, and excoriations; itching, burning, or stinging are common complains that may be experienced by the patients even in the absence of manifest dermatitis.[124]

The condition must be distinguished from other common skin diseases presenting with erythema in air-exposed areas, notably ACD, phototoxic and photoallergic dermatitis, atopic dermatitis, rosacea and seborrheic dermatitis, among others.

Irritant Dermatitis from Physical Factors

Physical factors at the workplace or in the household are important however often underestimated causes or culprits for skin irritancy.[131–133] The physical irritants most commonly implicated in the development of manifest dermatitis include friction, low ambient humidity/drying, heat, pressure, vibration, rubbing, and occlusion.[9,134]

Repeated skin damage through mechanical factors such as friction, rubbing, or pressure may induce localized forms of dermatitis that present with erythema, skin dryness, scaling, hyperkeratosis, or hyperpigmentation (friction dermatitis). Manual labor, personal protective equipment such as masks, helmets, or caps as well as medical devices (hearing aid, prostheses, continuous positive airway pressure masks) are common causative agents. Mechanical trauma may induce skin disease by Koebner phenomenon, facilitate the induction of contact sensitization, as well as play a role for development of acne mechanica or rare forms of physical urticaria.[135–137] Furthermore, prolonged mechanical stimulation through the use of body brushing utensils may lead to development of rippled-patterned brownish pigmentation referred to as friction melanosis.[138–145]

Low ambient humidity has been shown to be an independent risk factor for irritant dermatitis[146,147] and additionally, a common causative agent for dermatitis among workers in air-conditioned offices or cabin crew personnel.[134]

SUMMARY

ICD is a heterogeneous disease entity. Dermatitis or eczema presenting with erythema, vesicles, scaling, or fissures of different severity and course is the most common clinical form; however, the disease manifestations may extend beyond this pattern and result in noneczematous reactions such as urticaria, acne, folliculitis, hyper- or hypopigmentation, miliaria, granuloma formation, as well as hair or nail involvement. The diagnosis of ICD is based on the history and clinical findings supported by negative patch tests. In the absence of a routine diagnostic test, knowledge on the properties of the irritant, environmental, and host-related factors that influence the outcome of the irritant interaction with the skin may prove essential for correct and timely diagnosis. Increased awareness and advancement of knowledge on the pathogenesis of ICD are of primary importance for disease prevention at the workplace as well as in the general population.

REFERENCES

1. Dickel H, Kuss O, Schmidt A, et al. Importance of irritant contact dermatitis in occupational skin diseases. *Am J Clin Dermatol.* 2002;3:283–289.

2. Wigger-Alberti W, Elsner P. Contact Dermatitis due to irritation. In: Kanerva L, Elsner P, Wahlberg JE, Maibach HI, eds. *Handbook of Occupational Dermatology.* Berlin: Springer-Verlag; 2000:99–110.

3. Frosch PJ. Cutaneous irritation. In: Rycroft RJG, Menne T, Frosch PJ, eds. *Textbook of Contact Dermatitis.* Berlin: Springer; 1995:28–61.

4. Heinemann C, Paschold C, Fluhr J, et al. Induction of a hardening phenomenon by repeated application of SLS: analysis of lipid changes in the stratum corneum. *Acta Derm Venereol.* 2005;85:290–295.

5. Chew A, Maibach H. Ten genotypes of irritant contact dermatitis. In: Chew A, Maibach H, eds. *Irritant Dermatitis.* Berlin: Springer-Verlag; 2006:5–9.

6. Watkins SA, Maibach HI. The hardening phenomenon in irritant contact dermatitis: an interpretative update. *Contact Dermatitis.* 2009;60:123–130.

7. Malten KE. Thoughts on irritant contact dermatitis. *Contact Dermatitis.* 1981;7:238–247.

8. Weltfriend S, Ramon M, Maibach HI. Irritant dermatitis. In: Zhai H, Maibach HI, eds. *Dermatotoxicology.* Washington, DC: CRC Press; 2004:181–228.

9. Slodownik D, Lee A, Nixon R. Irritant contact dermatitis: a review. *Australas J Dermatol.* 2008;49:1–9.

10. Malten KE, den Arend JA, Wiggers RE. Delayed irritation: hexanediol diacrylate and butanedil diacrylate. *Contact Dermatitis.* 1979;5:178–184.

11. Lammintausta K, Maibach HI. Contact Dermatitis due to irritation: general principles, etiology and histology. In: Adams RM, ed. *Occupational Skin Disease*. Philadelphia, PA: WB Saunders Company; 1990:1–15.

12. Basketter DA, Marriott M, Gilmour NG, et al. Strong irritants masquerading as skin allergens: the case of benzalkonium chloride. *Contact Dermatitis*. 2004;5:213–217.

13. Uter W, Lessmann H, Geier J, et al. Is the irritant benzalkonium chloride a contact allergen? A contribution to the ongoing debate from a clinical perspective. *Contact Dermatitis*. 2008;58:359–363.

14. Fullerton A, Benfeldt E, Petersen JR, et al. The calcipotriol dose-irritation relationship: 48 hour occlusive testing in healthy volunteers using Finn Chambers. *Br J Dermatol*. 1998;138:259–265.

15. English JS. Current concepts of irritant contact dermatitis. *Occup Environ Med*. 2004;61:722–726,674.

16. Thyssen JP, Johansen JD, Linneberg A, et al. The epidemiology of hand eczema in the general population-prevalence and main findings. *Contact Dermatitis*. 2010;62:75–87.

17. Belsito DV. Occupational contact dermatitis: etiology, prevalence and resultant impairment/disability. *J Am Acad Dermatol*. 2005;53:303–313.

18. Diepgen TL. Occupational skin-disease data in Europe. *Int Arch Occup Environ Health*. 2003;76:331–338.

19. Chew AL, Maibach HI. Occupational issues of irritant contact dermatitis. *Int Arch Occup Environ Health*. 2003;76:339–346.

20. Kezic S, Visser M, Verberk M. Individual susceptibility to occupational contact dermatitis. *Industrial Health*. 2009;47:469–478.

21. Kezic S. Genetic susceptibility to occupational contact dermatitis. *Int J Immunopathol Pharmacol*. 2011;24:73S–78S.

22. Wilson A, de Vries N, Pociot F, et al. An allelic polymorphism within the human tumour necrosis factor alpha promoter region is strongly associated with HLA A1, B8 and DR3 allele. *J Exp Med*. 1993;177:557–560.

23. Wilson A, di Giovine FS, Blakemore AI, et al. Single base change in the tumour necrosis factor alpha (TNFA) gene detectable by NcoI restriction of PCR product. *Human Mol Genet*. 1992;1:353.

24. Allen M, Wakelin SH, Holloway D, et al. Association of TNFA gene polymorphism at position −308 with susceptibility to irritant contact dermatitis. *Immunogenetics*. 2000;51:201–205.

25. Davis JA, Visscher MO, Wickett RR, et al. Role of TNF-α polymorphism −308 in neurosensory irritation. *Int J Cosmet Sci*. 2011;33:105–112.

26. Landeck L, Visser M, Kezic S, et al. Impact of tumour necrosis factor-α polymorphisms on irritant contact dermatitis. *Contact Dermatitis*. 2012;66:221–227.

27. de Jongh CM, Khrenova L, Kezic S, et al. Polymorphisms in the interleukin-1 gene influence the stratum corneum interleukin-1 alpha concentration in uninvolved skin of patients with chronic irritant contact dermatitis. *Contact Dermatitis*. 2008;58:263–268.

28. Landeck L, Visser M, Kezic S, et al. IL1A-889 C/T gene polymorphism in irritant contact dermatitis. *J Eur Acad Dermatol Venereol*. 2013;27:1040–1043.

29. Brown SJ, McLean WH. One remarkable molecule: filaggrin. *J Invest Dermatol*. 2012;132:751–762.

30. Palmer CN, Irvine AD, Terron-Kwiatkowski A, et al. Common loss-of-function variants of the epidermal barrier protein filaggrin are a major predisposing factor for atopic dermatitis. *Nat Genet*. 2006;38:441–446.

31. Rodríguez E, Baurecht H, Herberich E, et al. Meta-analysis of filaggrin polymorphisms in eczema and asthma: robust risk factors in atopic disease. *J Allergy Clin Immunol.* 2009;123:1361–1370.

32. Van der Valk P, Nater J, Bleumink E. Vulnerability of the skin to surfactants in different groups of eczema patients and controls as measured by water vapour loss. *Clin Exp Dermatol.* 1985;10:98–103.

33. Agner T. Susceptibility of atopic dermatitis patients to irritant dermatitis caused by sodium lauryl sulphate. *Acta Dermatol Venereol.* 1991;71:296–300.

34. Tabata N, Tagami H, Kligman AM. A twenty-four hour occlusive exposure to 1% sodium lauryl sulfate induces a unique histopathologic inflammatory response in the xerotic skin of atopic dermatitis patients. *Acta Derm Venereol.* 1998;78:244–247.

35. Cowley NC, Farr PM. A dose–response study of irritant reactions to sodium lauryl sulphate in patients with seborrhoeic dermatitis and atopic eczema. *Acta Dermatol Venereol.* 1992;72:432–435.

36. Seidenari S. Reactivity to nickel sulfate at sodium lauryl sulfate pretreated skin sites is higher in atopics: an echographic evaluation by means of image analysis performed on 20 MHz B-scan recordings. *Acta Dermatol Venereol.* 1994;74:245–249.

37. Löffler H, Steffes A, Happle R, et al. Allergy and irritation: an adverse association in patients with atopic eczema. *Acta Dermatol Venereol.* 2003;83:328–331.

38. Jungersted JM, Scheer H, Mempel M, et al. Stratum corneum lipids, skin barrier function and filaggrin mutations in patients with atopic eczema. *Allergy.* 2010;65:911–918.

39. Angelova-Fischer I, Mannheimer AC, Hinder A, et al. Distinct barrier integrity phenotypes in filaggrin-related atopic eczema following sequential tape stripping and lipid profiling. *Exp Dermatol.* 2011;20:351–356.

40. Angelova-Fischer I, Dapic I, Hoek A, et al. Skin barrier integrity and natural moisturizing factor levels after cumulative dermal exposure to alkaline agents in atopic dermatitis. *Acta Dermatol Venereol.* 2013; (accepted).

41. John SM, Schwanitz IIJ. Functional skin testing: the SMART Procedures. In: Chew A, Maibach H, eds. *Irritant Dermatitis.* Berlin: Springer-Verlag; 2006:211–221.

42. Dickel H, Bruckner T, Schmidt A, et al. Impact of atopic skin diathesis on occupational skin disease incidence in a working population. *J Invest Dermatol.* 2003;121:37–40.

43. Lammintausta K, Kalimo K. Atopy and hand dermatitis in hospital wet work. *Contact Dermatitis.* 1981;7:301–308.

44. Meding B, Swanbeck G. Predictive factors for hand eczema. *Contact Dermatitis.* 1990;23:154–161.

45. Kristensen O. A prospective study of the development of hand eczema in an automobile manufacturing industry. *Contact Dermatitis.* 1992;26:341–345.

46. Rystedt I. Work-related hand eczema in atopics. *Contact Dermatitis.* 1985;12:164–171.

47. Bauer A, Bartsch R, Stadeler M, et al. Development of occupational skin disease during vocational training in baker and confectioner apprentices: a follow-up study. *Contact Dermatitis.* 1998;39:307–311.

48. Berndt U, Hinnen U, Iliev D, et al. Role of atopy score and of single atopic features as risk factors for the development of hand eczema in trainee metal workers. *Br J Dermatol.* 1999;140:922–924.

49. de Jongh CM, Khrenova L, Verberk MM, et al. Loss-of-function polymorphisms in the filaggrin gene are associated with an increased susceptibility to chronic irritant contact dermatitis: a case-control study. *Br J Dermatol.* 2008;159:621–627.

50. Visser MJ, Landeck L, Campbell LE, et al. Impact of atopic dermatitis and loss-of-function mutations in the filaggrin gene on the development of occupational irritant contact dermatitis. *Br J Dermatol*. 2013;168:326–332.

51. Thyssen JP, Carlsen BC, Menné T, et al. Filaggrin null mutations increase the risk and persistence of hand eczema in subjects with atopic dermatitis: results from a general population study. *Br J Dermatol*. 2010;163:115–120.

52. Landeck L, Visser M, Skudlik C, et al. Clinical course of occupational irritant contact dermatitis of the hands in relation to filaggrin genotype status and atopy. *Br J Dermatol*. 2012;167:1302–1309.

53. Willis CM. Variability in responsiveness to irritants: thoughts on possible underlying mechanisms. *Contact Dermatitis*. 2002;47:267–271.

54. Seyfarth F, Schliemann S, Antonov D, et al. Dry skin, barrier function, and irritant contact dermatitis in the elderly. *Clin Dermatol*. 2011;29:31–36.

55. Coenraads PJ, Bleumink E, Nater JP. Susceptibility to primary irritants: age dependence and relation to contact allergic reactions. *Contact Dermatitis*. 1975;1:377–381.

56. Schwindt DA, Wilhelm KP, Miller DL, et al. Cumulative irritation in older and younger skin: a comparison. *Acta Derm Venereol*. 1998;78:279–283.

57. Angelova-Fischer I, Becker V, Fischer TW, et al. Tandem repeated irritation in aged skin induces distinct barrier perturbation and cytokine profile in vivo. *Br J Dermatol*. 2012;167:787–793.

58. Elsner P, Wilhelm D, Maibach HI. Sodium lauryl sulphate-induced irritant contact dermatitis in vulvar and forearm skin of premenopausal and postmenopausal women. *J Am Acad Dermatol*. 1990;23:648–652.

59. Cua AB, Wilhelm KP, Maibach HI. Cutaneous sodium lauryl sulphate irritation potential: age and regional variability. *Br J Dermatol*. 1990;123:607–613.

60. Patil S, Maibach HI. Effect of age and sex on the elicitation of irritant contact dermatitis. *Contact Dermatitis*. 1994;30:257–264.

61. Robinson MK. Population differences in acute skin irritation responses. Race, sex, age, sensitive skin and repeat subject comparisons. *Contact Dermatitis*. 2002;46:86–93.

62. Meding B. Differences between the sexes with regard to work-related skin disease. *Contact Dermatitis*. 2000;43:65–71.

63. Agner T, Damm P, Skouby SO. Menstrual cycle and skin reactivity. *J Am Acad Dermatol*. 1991;24:566–570.

64. Altemus M, Rao B, Dhabhar FS, et al. Stress-induced changes in skin barrier function in healthy women. *J Invest Dermatol*. 2001;117:309–317.

65. de Jongh CM, Jakasa I, Verberk MM, et al. Variation in barrier impairment and inflammation of human skin as determined by sodium lauryl sulphate penetration rate. *Br J Dermatol*. 2006;154:651–657.

66. Feldmann RJ, Maibach HI. Regional variation in percutaneous penetration of 14C cortisol in man. *J Invest Dermatol*. 1967;48:181–183.

67. Elias PM, Cooper ER, Korc A, et al. Percutaneous transport in relation to stratum corneum structure and lipid composition. *J Invest Dermatol*. 1981;76:297–301.

68. Solomon AE, Lowe NJ. Percutaneous absorption in experimental epidermal disease. *Br J Dermatol*. 1979;100:717–722.

69. Dugard PH. Skin permeability theory in relation to measurements of percutaneous absorption in toxicology. *Adv Mod Toxicol*. 1977;4:525–550.

70. Fluhr JW, Dickel H, Kuss O, et al. Impact of anatomical location on barrier recovery, surface pH and stratum corneum hydration after acute barrier disruption. *Br J Dermatol.* 2002;146:770–776.

71. Ohlenschlaeger J, Friberg J, Ramsing D, et al. Temperature dependency of skin susceptibility to water and detergents. *Acta Derm Venerol.* 1996;76:274–276.

72. Clarys P, Manou I, Barel AO. Influence of temperature on irritation in the hand/forearm immersion test. *Contact Dermatitis.* 1997;36:240–243.

73. Fluhr JW, Praessler J, Akengin A, et al. Air flow at different temperatures increases sodium lauryl sulphate-induced barrier disruption and irritation in vivo. *Br J Dermatol.* 2005;152:1228–1234.

74. Agner T, Serup J. Seasonal variation in skin resistance to irritants. *Br J Dermatol.* 1989;121:323–328.

75. Cortesi PA, Scalone L, Belisari A, et al. Cost and quality of life in patients with severe chronic hand eczema refractory to standard therapy with topical potent corticosteroids. *Contact Dermatitis.* 2013; [Epub ahead of print].

76. Lindberg M, Bingefors K, Meding B, et al. Hand eczema and health-related quality of life; a comparison of EQ-5D and the Dermatology Life Quality Index (DLQI) in relation to the Hand Eczema Extent Score (HEES). *Contact Dermatitis.* 2013;69:138–143.

77. Moberg C, Alderling M, Meding B. Hand eczema and quality of life: a population-based study. *Br J Dermatol.* 2009;161:397–403.

78. Bruze M, Fregert S, Gruvberger B. Chemical skin burns. In: Kanerva L, Elsner P, Wahlberg JE, Meibach HI, eds. *Handbook of Occupational Dermatology.* Berlin: Springer-Verlag; 2000:325–332.

79. Bruze M, Gruvberger B, Fregert S. Chemical skin burns. In: Chew A, Maibach H, eds. *Irritant Dermatitis.* Berlin: Springer-Verlag; 2006:53–61.

80. Wilkinson SM, Beck MH. Contact dermatitis: irritant. In: Burns DA, Breatthnach SM, Cox NH, Griffiths CEM, eds. *Rook's Textbook of Dermatology 8th Edition.* 25. Chichester: Blackwell Publishing; 2010:1.

81. Kirkpatrick JJ, Enion DS, Burd DA. Hydrofluoric acid burns: a review. *Burns.* 1995;21:483–493.

82. Frosch PJ, John SM. Clinical aspects of irritant contact dermatitis. In: Johansen JD, et al., eds. *Contact Dermatitis.* Berlin: Springer-Verlag; 2011:305–345.

83. Adams RM. Cement burns. In: Adams RM, ed. *Occupational Skin Disease.* Philadelphia, PA: WB Saunders Company; 1990:15–16.

84. Spoo J, Elsner P. Cement burns: a review 1960–2000. *Contact Dermatitis.* 2001;45:68–71.

85. Poupon M, Caye N, Duteille F, et al. Cement burns: retrospective study of 18 cases and review of the literature. *Burns.* 2005;31:910–914.

86. Lewis PM, Ennis O, Kashif A, et al. Wet cement remains a poorly recognised cause of full-thickness skin burns. *Injury.* 2004;35:982–985.

87. Early SH, Simpson RL. Caustic burns from contact with wet cement. *JAMA.* 1985;254:528–529.

88. Fisher AA. Chromate dermatitis and cement burns. In: Fisher AA, ed. *Contact Dermatitis.* Philadelphia, PA: Lea & Febiger; 1986:762–772.

89. Lane PR, Hogan DJ. Chronic pain and scarring from cement burns. *Arch Dermatol.* 1985;121:368–369.

90. McGeown G. Cement burns of the hands. *Contact Dermatitis.* 1984;10:246.

91. Morley SE, Humzah D, McGregor JC, et al. Cement-related burns. *Burns.* 1996;22: 646–647.

92. Onuba O, Essiet A. Cement burns of the heels. *Contact Dermatitis.* 1986;14:325–326.

93. Stoermer D, Wolz G. Cement burns. *Contact Dermatitis.* 1983;9:421–422.

94. Tosti A, Peluso AM, Varotti C. Skin burns due to transit-mixed Portland cement. *Contact Dermatitis.* 1989;21:58.

95. Winder C, Carmody M. The dermal toxicity of cement. *Toxicol Ind Health.* 2002;18:321–331.

96. Kanerva L, Tarvainen K, Pinola A, et al. A single accidental exposure may result in a chemical burn, primary sensitization and allergic contact dermatitis. *Contact Dermatitis.* 1994;31:229–235.

97. Primka EJ, Taylor JS. Three cases of contact allergy after chemical burns from methylchloroisothiazolinone/methylisothiazolinone: one with concomitant allergy to methyldibromoglutaronitrile/phenoxyethanol. *Am J Contact Dermat.* 1997;8:43–46.

98. Isaksson M, Gruvberger B, Bruze M. Occupational contact allergy and dermatitis from methylisothiazolinone after contact with wallcovering glue and after a chemical burn from a biocide. *Dermatitis.* 2004;15:201–205.

99. McDonnell JK, Taylor JS. Occupational and environmental acne. In: Kanerva L, Elsner P, Wahlberg JE, Meibach HI, eds. *Handbook of Occupational Dermatology*. Berlin: Springer-Verlag; 2000:225–233.

100. O'Donovan WJ. Case of tar acne. *Proc R Soc Med.* 1922;15:50.

101. Adams BB, Chetty VB, Mutasim DF. Periorbital comedones and their relationship to pitch tar: a cross-sectional analysis and a review of the literature. *J Am Acad Dermatol.* 2000;42:624–627.

102. Plewig G, Fulton JE, Kligman AM. Pomade acne. *Arch Dermatol.* 1970;101:580–584.

103. Kligman AM, Mills OH. Acne cosmetica. *Arch Dermatol.* 1972;106:843–850.

104. Mills OH, Kligman A. Acne mechanica. *Arch Dermatol.* 1975;111:481–483.

105. Kokelj F. Occupational acne. *Clin Dermatol.* 1992;10:213–217.

106. Crow KD, Madli Puhvel S. Chloracne (halogen acne). In: Marzulli FN, Maibach HI, eds. *Dermatotoxicology*. New York, NY: Hemisphere; 1991:647–667.

107. Poland A, Glover E. Chlorinated biphenyl induction of aryl hydrocarbon hydroxylase activity: a study of the structure activity relationships. *Mol Pharmacol.* 1977;13:924–938.

108. Moore J.A. Toxicity of 2,3,7,8-tetrachlorodibenzo-*p*-dioxin. In: Ramel C, ed. Chlorinated phenoxy acids and their dioxins. *Ecol Bull.* 1978;27:134–144.

109. Goldmann PJ. Severe acute chloracne, a mass intoxication by 2,3,6,7-tetrachlorodibenzodioxin. *Hautarzt.* 1973;24:149–152.

110. Geusau A, Jurecka W, Nahavandi H, et al. Punctate keratoderma-like lesions on the palms and soles in a patient with chloracne: a new clinical manifestation of dioxin intoxication?. *Br J Dermatol.* 2000;143:1067–1071.

111. Suskind RR. Chloracne, the hallmark of dioxin intoxication. *Scand J Work Environ Health.* 1985;11:165–171.

112. Baccarelli A, Pesatori AC, Consonni D, et al. Health status and plasma dioxin levels in chloracne cases 20 years after the Seveso, Italy accident. *Br J Dermatol.* 2005;152:459–465.

113. Zober A., Ott M.G., Messerer P. Morbidity follow up study of BASF employees exposed to 2,3,7,8-tetrachlorodibenzo-*p*-dioxin (TCDD) after a 1953 chemical reactor incident. *Occup Environ Med.* 1994;51:479–486.

114. Geusau A, Abraham K, Geissler K, et al. Severe 2,3,7,8-tetrachlorodibenzo-*p*-dioxin (TCDD) intoxication: clinical and laboratory effects. *Environ Health Perspect.* 2001;109:865–869.

115. Farage M, Katsarou A, Maibach HI. Sensory, clinical and physiological factors in sensitive skin: a review. *Contact Dermatitis.* 2006;55:1–14.

116. Farage M, Maibach HI. Sensitive skin: closing in on a physiological cause. *Contact Dermatitis.* 2010;62:137–149.

117. Kligman AM, Sadiq I, Zhen Y, et al. Experimental studies on the nature of sensitive skin. *Skin Research and Technology.* 2006;12:217–222.

118. Ständer S, Schneider S, Weishaupt C, et al. Misery L. Putative neuronal mechanisms of sensitive skin. *Exp Dermatol.* 2009;18:417–423.

119. Denda M, Nakatani M, Ikeyama K, et al. Epidermal keratinocytes as the forefront of the sensory system. *Exp Dermatol.* 2007;16:157–161.

120. Frosch P, Kligman AM. Method for appraising the sting capacity of topically applied substances. *J Soc Cosmetic Chem.* 1977;28:197–209.

121. Marriott M, Holmes J, Peters L, et al. The complex problem of sensitive skin. *Contact Dermatitis.* 2005;53:93–99.

122. Schliemann S, Antonov D, Manegold N, et al. Sensory irritation caused by two organic solvents-short-time single application and repeated occlusive test in stingers and non-stingers. *Contact Dermatitis.* 2011;65:107–114.

123. Schliemann S, Antonov D, Manegold N, et al. The lactic acid stinging test predicts susceptibility to cumulative irritation caused by two lipophilic irritants. *Contact Dermatitis.* 2010;63:347–356.

124. Lachapelle JM. Airborne irritant dermatitis. In: Chew A, Maibach H, eds. *Irritant Dermatitis.* Berlin: Springer-Verlag; 2006:71–79.

125. Lachapelle JM. Airborne contact dermatitis. In: Rustemeyer T, Elsner P, John SM, Maibach HI, eds. *Kanerva's Occupational Dermatology.* Berlin: Springer-Verlag; 2012: 175–184.

126. Lachapelle JM. Industrial airborne irritant or allergic contact dermatitis. *Contact Dermatitis.* 1986;14:137–145.

127. Dooms-Goossens A, Deleu H. Airborne contact dermatitis: an update. *Contact Dermatitis.* 1991;25:211–217.

128. Santos R, Goossens A. An update on airborne contact dermatitis: 2001–2006. *Contact Dermatitis.* 2007;57:353–360.

129. Swinnen I, Goossens A. An update on airborne contact dermatitis: 2007–2011. *Contact Dermatitis.* 2013;68:232–238.

130. Lachapelle JM. Occupational air-borne skin diseases. In: Kanerva L, Elsner P, Wahlberg JE, Meibach HI, eds. *Handbook of Occupational Dermatology.* Berlin: Springer-Verlag; 2000:122–134.

131. Fluhr JW, Akengin A, Bornkessel A, et al. Additive impairment of the barrier function by mechanical irritation, occlusion and sodium lauryl sulphate in vivo. *Br J Dermatol.* 2005;153:125–131.

132. Kartono F, Maibach HI. Irritants in combination with a synergistic or additive effect on the skin response: an overview of tandem irritation studies. *Contact Dermatitis.* 2006;54:303–312.

133. McMullen E, Gawkrodger DJ. Physical friction is under-recognised as an irritant that can cause or contribute to contact dermatitis. *Br J Dermatol.* 2006;154:154–156.

134. Morris-Jones R, Robertson SJ, Ross JS, et al. Dermatitis caused by physical factors. *Br J Dermatol*. 2002;147:270–275.

135. Kennedy CTC, Burd DAR, Creamer D. Mechanical and thermal injury. In: Burns T, Breatnach S, Cox N, Griffiths C, eds. *Rook's Textbook of Dermatology*. Oxford, UK: Blackwell Publishing Ltd;; 2010:28.13–28.15.

136. Meneghini CL. Sensitization in traumatized skin. *Am J Int Med*. 1985;8:319–321.

137. Fischer T, Rystedt I. Cobalt allergy in hard metal workers. *Contact Dermatitis*. 1983;9:115–121.

138. Hayakawa R, Kato Y, Sugiura M, et al. Friction melanosis. In: Chew A, Maibach H, eds. *Irritant Dermatitis*. Berlin: Springer-Verlag; 2006:31–35.

139. Siragusa M, Ferri R, Cavallari V, et al. Friction melanosis, friction amyloidosis, macular amyloidosis, towel melanosis: many names for the same clinical entity. *Eur J Dermatol*. 2001;11:545–548.

140. Wong CK, Lin CS. Friction amyloidosis. *Int J Dermatol*. 1988;27:302–307.

141. Macsween RM, Saihan EM. Nylon cloth macular amyloidosis. *Clin Exp Dermatol*. 1997;22:28–29.

142. Hayakawa R, Suzuki M, Matsunnaga K. Friction melanosis. *Annu Rep Nagoya Univ Br Hosp*. 1991;25:32–38.

143. Hidano S. Friction melanosis. *Rinsho Derma*. 1984;26:1296–1297.

144. Kang HY, Rhee SH, Kim YC, et al. Friction melanosis and striae distensa caused by stretch training on a bench press. *J Dermatol*. 2005;32:765–766.

145. Iwasaki K, Mihara M, Nishiura S, et al. Biphasic amyloidosis arising from friction melanosis. *J Dermatol*. 1991;18:86–91.

146. Uter W, Gefeller O, Schwanitz HJ. An epidemiological study of the influence of season (cold and dry air) on the occurrence of irritant skin changes on the hands. *Br J Dermatol*. 1998;138:266–272.

147. Rycroft RJ. Environmental aspects of occupational dermatology. *Derm Beruf Umwelt*. 1986;34:157–159.

Toxicology

Golara Honari and Howard Maibach

INTRODUCTION

Skin irritations are complex biological phenomena, ranging from acute reactions following immediate contact to chronic dermatitis.[1,2] Chemicals can permeate through skin via intercellular lipid, trans cellular with direct permeation through the cornified cells, diffusion along hair follicle and sweat glands, or through traumatized skin.[3,4] Certain chemicals can cause severe tissue damage and necrosis, causing irreversible tissue damage, these are known as "corrosives." On the other hand "irritants" are substances that have reversible effects on skin; "acute irritants" cause inflammation after a single application, while "cumulative irritants" trigger irritation with recurrent exposure.[5] Repetitive exposures to substances with little intrinsic hazardous properties (such as water) are one of the leading causes of irritant dermatitis. Skin corrosion/irritation categories according to the United Nations (UN) Globally Harmonized System of Classification and Labeling of Chemicals (GHS)[6,7] are presented in Table 2.1. Optimal assessment of irritancy potential is best achieved when using the appropriate method for the irritation type. This chapter overviews methods of in vitro testing validated by European Centre for the Validation of Alternative Methods (ECVAM) to identify corrosives and differentiate them from noncorrosive substances as well as in vivo methods of corrosion/irritancy evaluation. The test guidelines published the Organization for Economic Co-operation and Development (OECD) are internationally agreed methods used to identify and characterize hazardous properties of new and existing chemicals.

PREDICTIVE IN VITRO TESTING

Corrosion Testing

The main principle of these assays is based on the fact that corrosive chemicals damage the stratum corneum and barrier function. Three

Table 2.1 Skin Irritation/Corrosion Categories by the Globally Harmonized System of Classification and Labeling of Chemicals (GHS)[6,7]

Skin Corrosion			Skin Irritation	Mild Skin Irritation
Category 1			Category 2	Category 3
Destruction of dermal tissue: visible necrosis in at least one of three animals			Reversible adverse effects in dermal tissue	Reversible adverse effects in dermal tissue
Subcategory 1A Exposure ≤ 3 min Observation ≤ 1 h	Subcategory 1B Exposure > 3 min ≤ 1 h Observation ≤ 14 days	Subcategory 1C Exposure > 1 h ≤ 4 h Observation ≤ 14 days	Draize score: ≥ 2.3 < 4.0 or persistent inflammation till day 14 in at least two of three tested animals	Draize score: ≥ 1.5 < 2.3 in at least two of three tested animals

in vitro methods validated by ECVAM are outlined below. In addition to the guideline 431 (originally adopted in 2004)[8], two other in vitro test methods for testing of corrosivity have been validated and adopted as OECD Test Guidelines 430[9] and 435.[10]

Assays that are currently validated for corrosion and irritation testing are listed in Table 2.2 and methods are outlined below.

Transcutaneous Electrical Resistance Test Method
This utilizes excised rat skin, and identifies corrosive materials by ability to damage barrier function, via measuring reduction in transcutaneous electrical resistance (TER).

The test material (150 μL for liquids or sufficient amount of a solid to evenly cover the tested skin and 150 μL of deionized water added on the top) is applied for up to 24 h to the epidermal surfaces of skin disks (three skin disks are used for each test and control substance) in a two-compartment test system. Skin disks function as the separation between the compartments. Measurements of electrical resistance is the primary endpoint, a reduction in the TER below a threshold level (5 kΩ for rat) indicates corrosivity.[11] Detailed guidelines are available at OECD Test Guideline 430; 2013.[9,12]

Membrane Barrier Test Method for Skin Corrosion
This utilizes an artificial membrane designed to respond to corrosive substances and may be used to test solids, liquids (with the exception

Table 2.2 European Union Reference Laboratory for Alternatives to Animal Testing (EURL-ECVAM) Validated In Vitro Test Methods for Skin Irritation/Corrosion Testing[31]

Test Method Name	Corrosion Testing	Irritation Testing
Transcutaneous electrical resistance test (TER)	Distinguish between corrosive and noncorrosive chemicals of different physical forms	N/A
Corrositex	Identification of corrosive properties for acids, bases, and their derivative	N/A
EpiSkin	Human skin model	Human skin model
SkinEthic	Human skin model	Human skin model
EpiDerm	Human skin model	Human skin model
epiCS	Human skin model	Human skin model

of aqueous materials with a pH between 4.5 and 8.5), and emulsions. Corrosive properties of a material are assessed based on time required for the tested material to penetrate through this bio-barrier (a proteinaceous gel, composed of protein, e.g., keratin, collagen, or mixtures of proteins) and a supporting filter membrane. The system is composed of a synthetic macromolecular bio-barrier and a Chemical Detection System (CDS), which can detect test substance. Tested material (500 µL of a liquid or 500 mg of a finely powdered solid) is applied evenly on the surface of the membrane barrier. Typically four replicas are produced for each tested substance and its corresponding controls. Controls include a noncorrosive vehicle or solvent used with the test substance, a positive control, typically a chemical with intermediate corrosivity such as sodium hydroxide (GHS Corrosive subcategory 1B)[6] and a negative (noncorrosive) control substance, such as10% citric acid or 6% propionic acid.[10] Corrositex™ is a commercially available membrane barrier test, validated by ECVAM. Detailed guidelines are available at OECD Test Guideline 435; 2006.[10]

Reconstructed Human Epidermis Test Method

Reconstructed human epidermis (RhE) models are three-dimensional biostructures composed of cultured normal human keratinocytes, which form a multilayered epidermis including stratum corneum. Assessments of skin corrosivity and/or irritancy via RhE models are based on the premise that corrosive substances penetrate through the stratum corneum and are toxic to the underlying cells. Corrosive chemicals are identified by their ability to decrease cell viability below defined threshold levels.[8]

Cell viability is measured by enzymatic conversion of a vital dye into a blue formazan salt that is quantitatively measured after extraction from tissues.

Two or three replicates are used for each test chemical and for the controls in each experiment. Sufficient amount of test chemical should be applied to uniformly cover the epidermis surface, exposure times, the incubation temperature, and applied chemical dose vary based on protocol details at OECD Test Guideline 431; 2013.[8] At the end of the exposure period, test chemical is carefully washed from the epidermis surface with aqueous buffer, or 0.9% NaCl. Then cell viability in RhE models is measured by enzymatic conversion of the vital dye MTT [3-(4,5-dimethylthiazol-2-yl)-2,5-diphenyltetrazolium bromide, thiazolyl blue], into a blue formazan salt that is quantitatively measured after extraction from tissues. Cell viability values distinguishing corrosive from noncorrosives vary and are listed in Table 2.3. EpiSkin®, EpiDerm®, and SkinEthic® test methods allow subcategorization of corrosive substances into category 1A and category 1B and 1C, in accordance with the UN GHS[6], but unable to distinguish between category 1B and category 1C (due to limited set of known category 1C chemicals). epiCS® test method is only used to identify corrosive versus noncorrosives.[8]

Irritation Testing
In Vitro RhE Test Method
Assessments of skin irritancy via RhE models are performed using basically the same test methods used to assess corrosivity, but with different exposure protocols.[8,13] Chemicals that lead to cell viability ≤ 50% are considered irritant (for UN GHS category 2)[7] and those with cell viability > 50% considered nonirritants. Protocol details are available at OECD Test Guideline 439; 2013.[13]

PREDICTIVE IN VIVO TESTING
Acute Dermal Irritation/Corrosion Testing
Assessment of skin corrosivity involving laboratory animals was adopted on 1981 and revised in 1992 and 2002, OECD Test Guideline 404.[14] This method provides information on corrosive potentials of a liquid or solid. Preferred sequential testing strategy is to perform validated in vitro testing prior to this test. New substances are initially

Test Method Name	Viability Measured After Exposure Time	Prediction to Be Considered
Table 2.3 The Prediction Model for Skin Corrosion Model[8]		
EpiSkin	<35% after 3 min exposure	Corrosive:
		• Optional subcategory 1A
	≥35% after 3 min exposure AND <35% after 60 min exposure OR ≥35% after 60 min exposure AND <35% after 240 min exposure	Corrosive: • Optional subcategory 1B and 1C
	≥35% after 240 min exposure	Noncorrosive
EpiDerm SCT and SkinEthic	<50% after 3 min exposure	Corrosive:
		• Optional subcategory 1A
	≥50% after 3 min exposure AND <15% after 60 min exposure	Corrosive: • Optional subcategory 1B and 1C
	≥50% after 3 min exposure AND ≥15% after 60 min exposure	Noncorrosive
epiCS	<50% after 3 min exposure OR (≥50% after 3 min exposure AND <15% after 60 min exposure)	Corrosive: Category 1
	≥50% after 3 min exposure AND >15% after 60 min exposure	Noncorrosive

evaluated via in vitro testing. In vivo testing is used when data is insufficient on dermal corrosion/irritation potentials of existing substances.

The test is based on evaluation of skin reactions to a single-dose application of a test substance in an experimental animal. Degree of irritation/corrosion is scored at specified intervals for complete evaluation of the effects. Duration of the study should be sufficient to evaluate the reversibility or irreversibility of the effects.

Initial Test (*In Vivo* Dermal Irritation/Corrosion)
In this method, albino rabbit is the preferred animal. Fur of the dorsal area is removed by close clipping 24 h prior to testing. The test substance is applied in a single dose to a small area of skin (~ 6 cm^2) of one animal and covered with a gauze patch and a nonirritating tape. Untreated skin areas serve as the control. Substances with a pH <2.0 or >11.5 are suspected to be corrosive and should not be tested. The dose is 0.5 mL (liquid) or 0.5 g (solid) applied to the test site and

covered with a gauze patch. The first patch is removed after 3 min. If no serious skin reaction is observed, a second patch will be placed on a different site for 1 h. A third patch will be applied for 4 h if the second patch is tolerated. After removal of the third patch, the response is evaluated according to the grading in Table 2.4. Initial testing is done using one animal. Animals should be examined for signs of erythema and edema, immediately after patch removal, at 60 min, and then at 24, 48, and 72 h.

If a corrosive effect is observed after any of the above sequential exposures, the test is immediately terminated. The animal will be observed for 14 days.

Confirmatory Test (*In Vivo* Dermal Irritation)

Confirmatory testing will be performed only in substances with no corrosive effect in initial testing. Two additional animals are tested, each with one patch of the test substance for 4 h. Animals should be examined for signs of erythema and edema at 60 min, and then at 24, 48, and 72 h, and for 14 days. Substance is considered irritant if inflammation persists by the end of this observation period. Additional animals may need to be used to clarify equivocal responses. Protocol details are available at OECD Test Guideline 404; 2002.[14]

Table 2.4 Grading of Dermal Responses (Draize Scale)[14,19]	
Erythema and Eschar Formation	
No erythema	0
Very slight erythema (barely perceptible)	1
Well-defined erythema	2
Moderate to severe erythema	3
Severe erythema (beef redness) to eschar formation	4
Edema Formation	
No edema	0
Very slight edema (barely perceptible)	1
Slight edema (edges of area well defined by definite raising)	2
Moderate edema (raised ~ 1 mm)	3
Severe edema (raised >1 mm and extending beyond area of exposure)	4

Note: *The primary irritation index (PII) is calculated by adding up the average of the erythema and edema values.*
PII < 2: mildly irritating
2 < PII < 5: moderately irritating
PII > 5: severely irritating (require precautionary labeling)

The Draize Rabbit Skin Irritancy Test

The Draize rabbit skin used successfully since 1940s[15] and established the bases for the above methods used in the test guideline 404, involves application of two semioccluded patches of an undiluted chemical to the shaved back of each animal (typically three albino rabbits). Each material is tested on two 1-inch-square sites on the same animal; one site is intact and one is abraded in such a way that the stratum corneum is opened but no bleeding produced. Each test site is covered with two layers of 1-inch-square surgical gauze secured in place with tape. Patches are removed 24 h after application, and the test sites evaluated for erythema and edema using a prescribed scale with the degree of irritation scored for erythema (redness), eschar (scab formation), edema (swelling), and corrosive action at 24, 48, and 72 h (Table 2.4).[5] Other animal assays assess cumulative irritation caused by some chemicals 230 which involves both exaggeration and repetition of exposures such as guinea pig immersion test, or repeated patch test in a modified rabbit test.[16] These tests are not required by regulatory agencies. However, in most instances, these methods have not gained widespread acceptance.

Currently, internationally accepted test methods for skin corrosion and irritation testing include the traditional in vivo animal test (based on Draize rabbit test) as well as in vitro test methods (discussed earlier). Animal testing for cosmetic products has been prohibited in Europe. Testing ban and marketing ban of cosmetic products, which were tested on animals, has been in effect since March 2013.[17]

Human Irritation Testing

Controlled studies with human volunteers have provided meaningful data and their basic principles are discussed below. Predicative irritation assays in human, using a small area on skin, can be done provided that systematic toxicity (from absorption) is low and informed consent obtained. Although regulatory agencies do not routinely require testing in humans, human tests are sometimes preferred to animal tests. Test sites generally heal rapidly, within a week or so. More severe reactions should be evaluated periodically over a longer time to ensure resolution and determine inflammatory patterns. Some subjects may develop changes in pigmentation level at the test site following severe responses. Detailed consultation before consenting human subjects are extremely important.

SINGLE-APPLICATION IRRITATION PATCH TEST

This is used for assessment of acute irritation potential and involves application of 0.5 mL (0.5 g for solid test materials) on a 25 mm chamber (0.2 mL and 0.2 g have also been used)[18]; solid test materials are moistened, to the skin of human volunteers for up to 4 h. New test materials may be applied for shorter periods (30 min to 1 h) and if tolerated exposure time can progressively be increased to 4 h. When testing new materials, application of dilutions should be considered.[19] Subjects should routinely be instructed to remove patches immediately if unusual discomfort occurs. Tested sites are assessed for the presence of irritation using a grading scale (Table 2.5) at 24, 48, and 72 h after patch removal.[18] Commercial patches, chambers, gauze squares, or cotton bandage material can be used and patches are applied to either upper back or to the dorsal surface of the upper arm. Patches are secured with surgical tape without wrapping the trunk of the arm. For volatile materials, a relatively nonocclusive tape, such as Micropore®, or Scanpore®, should be used.[5] This test detects acute skin irritation hazard potential and should be used for hazard classification. It is not intended to predict other types of skin irritant dermatitis, such as cumulative irritant dermatitis.[20]

REPEAT-APPLICATION CUMULATIVE IRRITATION PATCH TEST

These assays assess cumulative irritation. The early work of Marzulli and Maibach and Kligman and Wooding and other investigators formed the bases of cumulative irritation assays.[16,19,21,22] In the original assay, test material was applied to a 25 mm^2 skin of back either via a saturated Webril™ or the skin surface was covered with a viscous material. Patches were removed after 24 h, sites evaluated and a fresh set of patches reapplied up to 21 days. Basic principals are similar to the single application and many investigators have developed their version of it. Shorter study times have also been used in evaluation of surfactants. Many investigators have developed their version.[5,21] These methods do not always predict safety of consumer products, which may be related to intrinsic differences in reactivity of healthy skin versus damaged or sensitive skin.[23]

EXAGGERATED EXPOSURE IRRITATION TESTS

Irritancy patch testing does not always correlate with the consumers' experience. Soaps are an example in which patch testing may overpredict the reaction in real life. Immersion testing and antecubital washing

Table 2.5 Human Patch Test Grading Scales[5,19]	
Detailed Human Patch Test Grading Scale	
No apparent cutaneous involvement	0
Faint, barely perceptible erythema, or slight dryness (glazed appearance)	1/2
Faint but definite erythema, no eruptions or broken skin *or* no erythema but definite dryness; may have epidermal fissuring	1
Well-defined erythema or faint erythema with definite dryness; may have epidural fissuring	1–1/2
Moderate erythema, may have a *few* papules or deep fissures, moderate-to-severe erythema in the cracks	2
Moderate erythema with barely perceptible edema *or* severe erythema not involving a significant portion of the patch (halo effect around the edges), may have a few papules *or* moderate to severe erythema	2–1/2
Severe erythema (beet red). May have generalized papules *or* moderate to severe erythema with slight edema (edges well defined by raising)	3
Moderate to severe erythema with moderate edema (confined to patch area) *or* moderate to severe erythema with isolated eschar formations or vesicles	3–1/2
Generalized vesicles *or* eschar formation *or* moderate to severe erythema and/or edema extending beyond patch area	4
Simple Patch Test Grading Scale	
Negative, normal skin	0
Questionable erythema not covering entire area	±
Definite erythema	1
Erythema and induration	2
Vesiculation	3
Bullous reaction	4

test, also known as flex wash, are examples of nonpatch irritancy techniques. Numerous versions of these tests have been used by investigators basically exposing skin of consented human subjects to different concentrations of detergents for varied amounts of time, endpoints are erythema and scaling.[19,24,25]

USE OF BIOENGINEERING DEVICES

Multiple commercially available devices are used to measure biophysical properties of skin. These measures provide objective data to be used in conjunction with clinical evaluation of inflammatory responses (detailed discussion in Chapter 1).

- Transepidermal water loss (TEWL) measurements: TEWL reflects the integrity of the barrier function and can be determined by the use of evaporimeter.[26]

- Laser Doppler velocimetry: This has been used to quantify the increased blood flow to inflamed tissue and can be used for estimation of microcirculation[27]
- Squamometry: This is a noninvasive, protein-dependent, colorimetric evaluation of the stratum corneum and is a sensitive tool in the assessment of nonerythematous irritant dermatitis. An adhesive disk (D-SQUAME) is applied on the skin and upon removal of the tape superficial desquamating layer of the stratum corneum is harvested. Disks are subsequently stained and the amount dye found in the cells are quantified and scaled.[28–30]

REFERENCES

1. Malten KE, den Arend JA. Irritant contact dermatitis. traumiterative and cumulative impairment by cosmetics, climate, and other daily loads. *Derm Beruf Umwelt*. 1985;33(4): 125–132.

2. Chew A, Maibach HI. Ten genotypes of irritant contact dermatitis. In: Chew A, Maibach HI, eds. *Irritant Dermatitis*. Germany: Springer; 2006:5.

3. Moser K, Kriwet K, Naik A, Kalia YN, Guy RH. Passive skin penetration enhancement and its quantification in vitro. *Eur J Pharm Biopharm*. 2001;52(2):103–112.

4. Elias PM. The epidermal permeability barrier: from the early days at Harvard to emerging concepts. *J Invest Dermatol*. 2004;122(2):xxxvi–xxxix.

5. Hayes BB, Patrick E, Maibach HI. Dermatotoxicology. In: Hayes W, ed. *Principles and Methods of Toxicology, Fifth Edition*. 5th ed. Canada: CRC press; 2007:1359.

6. A Guide to the Globally Harmonized System of Classification and Labelling of Chemicals (GHS). <https://www.osha.gov/dsg/hazcom/ghs.html> Accessed 26.1.14.

7. UN-GHS. Health hazards-part 3. <http://www.unece.org/fileadmin/DAM/trans/danger/publi/ghs/ghs_rev03/English/03e_part3.pdf> Accessed 26.1.14.

8. OECD 431-2013. *In Vitro Skin Corrosion: Reconstructed Human Epidermis (RHE) Test Method*. <http://www.oecd-ilibrary.org/docserver/download/9713211e.pdf?expires=1400505488&id=id &accname=guest&checksum=E243CD11FEA17FA3AFD721AA57B5A2EF> Accessed 18.05.2014.

9. OECD, 430-2004. *In Vitro Skin Corrosion: Transcutaneous Electrical Resistance Test (TER)*, OECD guidelines for the testing of chemicals, section 4, OECD publishing. <http://www.oecd-ilibrary.org/content/book/9789264071124-en> Accessed 24.1.14.

10. OECD 435-2006. *OECD Guideline for the Testing of Chemicals In Vitro Membrane Barrier Test Method for Skin Corrosion—435*. <http://www.oecd-ilibrary.org/docserver/download/9743501e. pdf?expires=1400505414&id=id&accname=guest&checksum=E1597B77EB5F93FA5E446 BC20E98D6A5> Accessed 19.05.2014.

11. Oliver GJA, Pemberton MA, Rhodes C. An in vitro skin corrosivity test—modifications and validation. *Food Chem Toxicol*. 1986;24(6–7):507–512.

12. OECD, 430—2013. *In Vitro Skin Corrosion: Transcutaneous Electrical Resistance Test Method (TER)*.   <http://www.oecd-ilibrary.org/environment/test-no-430-in-vitro-skin-corrosion-transcutaneous-electrical-resistance-test-method-ter_9789264203808-en;jsessionid = 30c2gp32 t1sbg.x-oecd-live-02> Accessed 18.05.2014.

13. OECD, 439-2013. *OECD Guidelines for the Testing of Chemicals In vitro Skin irritation: Reconstructed human Epidermis Test method—439*. <http://ntp.niehs.nih.gov/iccvam/SuppDocs/FedDocs/OECD/OECD-TG439-2013-508.pdf> Accessed 24.01.14.

14. Test no. 404: Acute dermal irritation/corrosion. <http://www.oecd-ilibrary.org/content/book/9789264070622-en> Accessed 22.10.13.

15. Draize JH, Woodard G, Calvery HO. Methods for the study of irritation and toxicity of substances applied topically to the skin and mucous membranes. *J Pharmacol Exp Ther.* 1944;82:377.

16. Marzulli FN, Maibach HI. The rabbit as a model for evaluating skin irritants: a comparison of results obtained on animals and man using repeated skin exposures. *Food Cosmet Toxicol.* 1975;13(5):533–540.

17. Joint Research Centre of The European Commission's in-house science service. *Advancing Safety Assessment without Animals: EURL ECVAM* <http://ihcp.jrc.ec.europa.eu/our_labs/eurl-ecvam/clip-dec-13/eurl-ecvam-script-Dec-2013.pdf> Accessed 5.01.14.

18. Basketter DA, York M, McFadden JP, Robinson MK. Determination of skin irritation potential in the human 4-h patch test. *Contact Dermatitis.* 2004;51(1):1–4.

19. Patil SM, Patrick E, Maibach HI. Animal, human, and in vitro test methods for predicting skin irritation. In: Marzulli FN, Maibach HI, eds. *Dermatotoxicology Methods.* Washington, DC, USA: CRC Press; 1998:89–114.

20. Jirova D, Basketter D, Liebsch M, et al. Comparison of human skin irritation patch test data with in vitro skin irritation assays and animal data. *Contact Dermatitis.* 2010;62(2):109–116.

21. Kligman AM, Wooding WM. A method for the measurement and evaluation of irritants on human skin. *J Invest Dermatol.* 1967;49(1):78–94.

22. Philips L, Steinberg M, Maibach HI. A comparison of rabbit and human skin response to certain irritants. *Toxicol Appl Pharmacol.* 1972;21(3):369–382.

23. Frosch PJ, Kligman AM. The soap chamber test. A new method for assessing the irritancy of soaps. *J Am Acad Dermatol.* 1979;1(1):35–41.

24. Kooyman DJ, Snyder FM. Tests for the mildness of soaps. *Arch Dermatol Syph.* 1942;46:846–855.

25. Gabard B, Chatelain E, Bieli E, Haas S. Surfactant irritation: in vitro corneosurfametry and in vivo bioengineering. *Skin Res Technol.* 2001;7(1):49–55.

26. Laudanska H, Reduta T, Szmitkowska D. Evaluation of skin barrier function in allergic contact dermatitis and atopic dermatitis using method of the continuous TEWL measurement. *Rocz Akad Med Bialymst.* 2003;48:123–127.

27. Bircher A, de Boer EM, Agner T, Wahlberg JE, Serup J. Guidelines for measurement of cutaneous blood flow by laser Doppler flowmetry. A report from the standardization group of the European Society of Contact Dermatitis. *Contact Dermatitis.* 1994;30(2):65–72.

28. Hendrix SW, Miller KH, Youket TE, et al. Optimization of the skin multiple analyte profile bioanalytical method for determination of skin biomarkers from D-squame tape samples. *Skin Res Technol.* 2007;13(3):330–342.

29. Black D, Boyer J, Lagarde JM. Image analysis of skin scaling using D-squame samplers: comparison with clinical scoring and use for assessing moisturizer efficacy. *Int J Cosmet Sci.* 2006;28(1):35–44.

30. Serup J, Winther A, Blichmann C. A simple method for the study of scale pattern and effects of a moisturizer—qualitative and quantitative evaluation by D-squame tape compared with parameters of epidermal hydration. *Clin Exp Dermatol.* 1989;14(4):277–282.

31. Institute for Health and Consumer Protection (IHCP). <http://ihcp.jrc.ec.europa.eu/our_labs/eurl-ecvam/validation-regulatory-acceptance/topical-toxicity/skin-irritation> Accessed 20.10.13.

CHAPTER 3

Photoirritation (Phototoxicity): Clinical Aspects

Howard Maibach and Golara Honari

Key Points
- **Exposure to ultraviolet (UV) radiation and visible light can precipitate toxic reactions referred to as phototoxic or photoirritant dermatitis.**
- **Phototoxicity reactions are evoked by the interaction of the above radiations with endogenous or exogenous compounds.**
- **Phototoxicity is a nonimmunologic process.**
- **Enhanced photocarcinogenesis.**

INTRODUCTION

Solar radiation, necessary for numerous physiologic processes in human skin, is a ubiquitous environmental stressor. Ultraviolet (UV), visible light, and infrared (IR), collectively referred to as optical radiation, are most relevant in human photobiology.

Physiologic and pathologic processes occur with acute and chronic exposure to UV radiation, including skin darkening, vitamin D production, sunburn, carcinogenesis, immunosuppression, and photodermatoses. Photodermatoses are diseases caused or aggravated by exposure to UV radiation or visible light.[1,2] While many photoactive chemicals can induce photosensitivity, multiple congenital or acquired conditions can make individuals more sensitive to light. Examples of these conditions include immunologically mediated disorders (i.e., polymorphic light eruption, actinic prurigo, hydroa vacciniforme, chronic actinic dermatitis, solar urticarial), genetic syndromes with defective DNA repair, photoaggravated dermatoses (i.e., lupus erythematosus), and endogenous chemical-induced photosensitivities (i.e., cutaneous porphyrias).[2]

This chapter discusses phototoxic reactions caused by exogenous systemic or topically applied drugs or chemicals.

Applied Dermatotoxicology. DOI: http://dx.doi.org/10.1016/B978-0-12-420130-9.00003-7

BASIC PRINCIPLES OF PHOTOBIOLOGY

Biological effects of solar radiation vary with wavelength. UV radiation is subdivided into UVA (400–320 nm), UVB (320–290 nm), and UVC (290–200 nm). Earth's atmosphere absorbs almost all the higher energy UVs, namely, UVC and much of UVB. Therefore, the UV reaching the earth surface is about 95% UVA and 5% UVB.[3] Photon energy penetrating skin is able to cause molecular reactions in the tissue. Wavelengths shorter than 320 nm are more biologically relevant; however, advances in molecular photobiology indicate a subdivision at 330–340 may be more appropriate, hence UVA has been subdivided into UVA1 (400–340 nm) and UVA2 (340–315 nm).[4] Approximate depth of penetration of various wavelengths in human skin is represented in Figure 3.1.

In tissue, numerous molecules and microscopic particles, known as chromophores, absorb UV radiation. Radiation energy initiates numerous thermochemical reactions, leading to molecular and cellular changes with consequent clinical implications. DNA is the most prominent chromophore in skin. Other endogenous chromophores such as melanin, hemoglobin, porphyrins exist in human skin.[5]

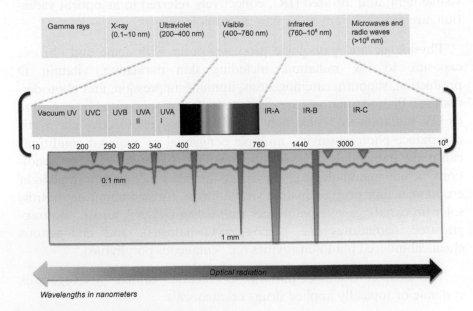

Figure 3.1 Electromagnetic spectrum and major wavelength regions and penetration.

Acute UV exposure can lead to sunburn, tanning, UV-induced immunosuppression, and DNA damage. Chronic UV exposure is associated with photoaging, cumulative DNA damage, and increased carcinogenesis.[6] Mechanisms of photocarcinogenesis are beyond the scope of this chapter. UVA penetrates the skin more deeply and is the most important spectrum for absorption by systemic chromophores. UVB is mostly absorbed by the epidermis and has been mostly associated with erythema,[7] photocarcinogenesis,[8] and rarely with exogenous photosensitivity.[9] While delayed pigmentation is caused by UVB,[10] immediate pigment darkening is more due to UVA.[11]

MECHANISMS OF PHOTOTOXIC REACTIONS

Phototoxic reactions are nonimmunologic reactions that are caused by interaction of UV radiation with an endogenous or exogenous chromophores (photoirritations). Immunologic UV-induced reactions (photoallergic reactions) are discussed in Chapter 5. Phototoxicity is the result of direct cellular damage following the exposure to UV radiation or visible light in the presence of a phototoxic substance. Phototoxic agents can be endogenous or exogenous. Unlike photoallergic conditions, which are immunologically mediated and only involve certain individuals, phototoxic reaction can essentially affect most individuals as long as they are exposed to sufficient amount of radiation in the presence of a photoirritant.[12]

Phototoxic reactions are initiated following absorption of UV radiation, a visible light photon by a chromophore or a photosensitizer. Energy promotes the electrons in the target molecule to reach a higher energy level, from a so-called ground state to an excited state. Each chromophore has a unique absorption spectrum. Excited molecules will return to ground state within a few microseconds (in triplet excited state) or within nanoseconds (in singlet excited state). As these molecules return to the ground states, they release light or heat or undergo a photochemical reaction, which can convert the chromophore into a new stable molecule called photoproduct that can initiate cellular changes and lead to clinical reactions.[4] An excited state chromophore may transfer its energy to oxygen. Reactive oxygen species (ROS) are small molecules and free radicals including singlet oxygen, hydrogen peroxide, superoxide anion, and hydroxyl radical. They can oxidize unsaturated lipids of cell membranes and organelles, aromatic amino acids, and pyrimidine bases of DNA or RNA. Oxidized products can trigger release of multiple inflammatory mediators that cause inflammation and erythema.[13]

Some phototoxic agents have also the capability to induce chromosomal damage with photomutagenic and photoimmunosuppressive properties with consequent carcinogenesis.[14,15] Increased incidence of skin cancer in patients on long-term PUVA treatments,[16] also in patients exposed to fluorquinolones,[17,18] diuretics,[19] and voriconazole[20] has been documented.

CLINICAL PRESENTATIONS

Phototoxic reactions typically resemble an acute sunburn reaction. Painful erythema of photo-exposed skin develops within minutes to hours of sun exposure. However, they can vary from urticarial, eczematous, lichenoid, or pigmentary changes. Clinical distinction between phototoxic and photoallergic reactions is difficult; also both conditions can exist concomitantly. Photoallergens and photoactive agents may have irritancy potentials as well. Key manifestations of photosensitivities are given in Table 3.1.

Table 3.1 Key Features of Photosensitivity Reactions

	Phototoxic Reactions	Photoallergic Reactions
Incidence	High	Low
Sensitization required	No	Yes
Clinical presentation	Sunburn-like reaction, erythema, edema with or without blistering, burning and painful, sharply demarcated	Eczematous, itchy
Onset of symptoms	Minutes to hours after photoexposure	24–48 h after photoexposure
Pathophysiology	Direct cell damage, DNA damage, ROS, inflammation	Type IV hypersensitivity photoproduct
Histology	Epidermal necrosis, dermal edema	Spongiotic dermatitis, dermal lymphohistiocytic infiltrate
Cross-reactivity	No	Often
Diagnosis	Clinical and phototests	Clinical and photopatch tests

Distribution of skin lesions in phototoxicity depends on whether the photoactive chemical is topically applied (photocontact dermatitis) or a systemic photosensitizer is the cause of reaction. In photocontact dermatitis from a topical agent, dermatitis draws the area of application and photoexposure. In systemic photosensitivity, the involved areas are typically symmetric, all exposed areas of the face, the V-shaped area of the neck, hand and forearms, sparing photo-protected skin, such as retroauricular and submandibular areas as shown in Figure 3.2. Involvement of these shaded areas suggests dermatitis from an airborne allergen or irritant.[21]

Timing of phototoxic reaction following an assault also varies. Symptoms can develop acutely after the exposure resembling urticarial reactions or acute sunburn. Subacute reactions are less frequent and may develop within days to weeks after exposure to the photosensitizer and the sun. Pseudoporphyria, photoonycholysis, hyper- or hypopigmentation, telangicctasia, and purpura are the subacute phototoxic reactions. Delayed manifestations of photosensitivity include photocarcinogenesis, subacute and chronic lupus erythematosus.[21]

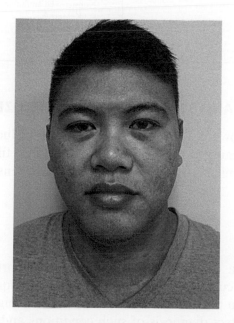

Figure 3.2 *Photo-induced dermatitis often sparing shaded areas of skin; including upper eyelids, nasolabial folds and submental areas.*[22]

Table 3.2 Common Phototoxic and Photoallergic Agents[2]

Common Phototoxic Agents	Common Photoallergic Agents

• Antiarrhythmics
  – Amiodarone
  – Quinidine
• Triazole antifungals
  – Voriconazole
• Diuretics
  – Furosemide
  – Thiazides
• Nonsteroidal antiinflammatory drugs
  – Nabumetone
  – Naproxen
  – Piroxicam
• Phenothiazines
  – Chlorpromazine
  – Prochlorperazine
• Psoralens
  – 5-Methoxypsoralen
  – 8-Methoxypsoralen
  – 4,5′,8-Trimethylpsoralen
• Quinolones
  – Ciprofloxacin
  – Lomefloxacin
  – Nalidixic acid
  – Sparfloxacin
• St. John's wort
  – Hypericin
• Sulfonamides
• Sulfonylureas
• Tar (topical)
• Tetracyclines
  – Doxycycline
  – Demeclocycline
	• Topical agents:
	  • Sunscreens (e.g., oxybenzone [benzophenone-3])
	  • Fragrances
	    – 6-Methylcoumarin
	    – Musk ambrette
	    – Sandalwood oil
	  • Antimicrobial agents
	    – Bithionol
	    – Chlorhexidine
	    – Fenticlor
	    – Hexachlorophene
	  • Nonsteroidal antiinflammatory drugs
	    – Diclofenac
	    – Ketoprofen
	  • Phenothiazines
	    – Chlorpromazine
	    – Promethazine
	• Systemic agents:
	  • Antiarrhythmics
	    – Quinidine
	  • Antimalarials
	    – Quinine
	  • Antifungals
	    – Griseofulvin
	  • Antimicrobials
	    – Quinolones (e.g., enoxacin, lomefloxacin)
	    – Sulfonamides
	  • Nonsteroidal antiinflammatory drugs
	    – Ketoprofen
	    – Piroxicam

MAIN TOPICAL AND SYSTEMIC PHOTOSENSITIZERS

A large and growing list of photosensitizers includes plant-driven photosensitizers causing phytophotodermatitis, UV filters, number of topical and systemic medications. Common photosensitizers are listed in Table 3.2.

SUMMARY

Phototoxicity is a common and possibly underreported polymorphic condition with acute and/or chronic clinical consequences. Multiple agents can lead to phototoxic reactions and it is of high importance to stay alert for correct diagnosis of such conditions and identification of new phototoxic agents (Figure 3.3).

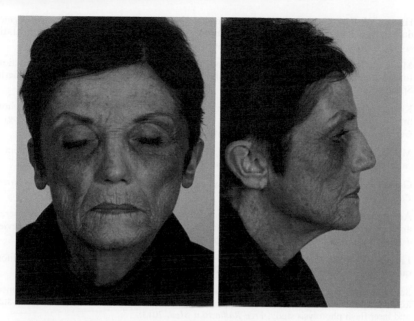

Figure 3.3 Phototoxic reaction caused by imipramine. Blue-gray discoloration on face has been present for almost 10 years with a 30 year history of imipramine ingestion. Photograph Courtesy of Dr Pooja Khera, MD. Cleveland Clinic, Cleveland. Ohio.[23]

REFERENCES

1. Morikawa F, Fukuda M, Naganuma M, Nakayama Y. Phototoxic reaction to xanthene dyes induced by visible light. *J Dermatol.* 1976;3(2):59−67.

2. Lim HW, Hawk JLM. Photodermatologic disorders. In: 3rd ed. Bolognia JL, Jorizzo JL, Schaffer JV, eds. *Dermatology.* vol. 2. China: Elsevier; 2012:1467−1486.

3. Polefka TG, Meyer TA, Agin PP, Bianchini RJ. Effects of solar radiation on the skin. *J Cosmet Dermatol.* 2012;11(2):134−143.

4. Diffey BL, Kochevar IE. Basic principles of photobiology. In: Lim HW, Honigsmnn H, Hawk JLM, eds. *Photodermatology.* USA; 2007:15−27.

5. Wondrak GT, Jacobson MK, Jacobson EL. Endogenous UVA-photosensitizers: mediators of skin photodamage and novel targets for skin photoprotection. *Photochem Photobiol Sci.* 2006;5(2):215−237.

6. Matsumura Y, Ananthaswamy HN. Toxic effects of ultraviolet radiation on the skin. *Toxicol Appl Pharmacol.* 2004;195(3):298−308.

7. Anders A, Altheide HJ, Knalmann M, Tronnier H. Action spectrum for erythema in humans investigated with dye lasers. *Photochem Photobiol.* 1995;61(2):200−205.

8. de Gruijl FR. Action spectrum for photocarcinogenesis. *Recent Results Cancer Res.* 1995;139:21−30.

9. Fujimoto N, Danno K, Wakabayashi M, Uenishi T, Tanaka T. Photosensitivity with eosino-philia due to ambroxol and UVB. *Contact Dermatitis.* 2009;60(2):110−113.

10. Parrish JA, Jaenicke KF, Anderson RR. Erythema and melanogenesis action spectra of normal human skin. *Photochem Photobiol.* 1982;36(2):187−191.

11. Hwang YJ, Park HJ, Hahn HJ, et al. Immediate pigment darkening and persistent pigment darkening as means of measuring the ultraviolet A protection factor in vivo: a comparative study. *Br J Dermatol.* 2011;164(6):1356–1361.

12. Mang R, Stege H, Krutmann J. Mechanisms of phototoxic and photoallergic reactions. In: Johansen JD, Frosch PJ, Lepoittevin JP, eds. *Contact Dermatitis.* 5th ed. Berlin: Springer; 2011:155–163.

13. Kochevar IE, Taylor CR, Krutmann J. Fundamentals of cutaneous photobiology and photoimmunology. In: Goldsmith LA, Katz SI, Gilchrest BA, eds. *Fitzpatrick's Dermatology in General Medicine.* 8th ed. USA: McGraw-Hill Professional; 2012.

14. Muller L, Kasper P, Kersten B, Zhang J. Photochemical genotoxicity and photochemical carcinogenesis—two sides of a coin? *Toxicol Lett.* 1998;102–103:383–387.

15. Gocke E. Photochemical mutagenesis: examples and toxicological relevance. *J Environ Pathol Toxicol Oncol.* 2001;20(4):285–292.

16. Archier E, Devaux S, Castela E, et al. Carcinogenic risks of psoralen UV-A therapy and narrowband UV-B therapy in chronic plaque psoriasis: a systematic literature review. *J Eur Acad Dermatol Venereol.* 2012;26(suppl 3):22–31.

17. Lhiaubet-Vallet V, Bosca F, Miranda MA. Photosensitized DNA damage: the case of fluoroquinolones. *Photochem Photobiol.* 2009;85(4):861–868.

18. Soldevila S, Consuelo Cuquerella M, Lhiaubet-Vallet V, Edge R, Bosca F. Seeking the mechanism responsible for fluoroquinolone photomutagenicity: a pulse radiolysis, steady-state, and laser flash photolysis study. *Free Radic Biol Med.* 2013;.

19. Jensen AO, Thomsen HF, Engebjerg MC, Olesen AB, Sorensen HT, Karagas MR. Use of photosensitising diuretics and risk of skin cancer: a population-based case–control study. *Br J Cancer.* 2008;99(9):1522–1528.

20. McCarthy KL, Playford EG, Looke DF, Whitby M. Severe photosensitivity causing multifocal squamous cell carcinomas secondary to prolonged voriconazole therapy. *Clin Infect Dis.* 2007;44(5):e55–e56.

21. Gonçalo M. Phototoxic and photoallergic reactions. In: Johansen JD, Frosch PJ, Lepoittevin JP, eds. *Contact Dermatitis.* 5th ed. Berlin, Heidelberg: Springer-Verlag; 2011:361–376.

22. Kaddu S, Kerl H, Wolf P. Accidental bullous phototoxic reactions to bergamot aromatherapy oil. *J Am Acad Dermatol.* 2001;45(3):458–461.

23. Susser WS, Whitaker-Worth DL, Grant-Kels JM. Mucocutaneous reactions to chemotherapy. *J Am Acad Dermatol.* 1999;40(3):367–398.

Phototoxicity Testing

Golara Honari and Howard Maibach

INTRODUCTION

Phototoxicity (photoirritation) occurs when certain chemicals in skin absorb light and create a pathological irritation. These chemicals may be applied topically or diffuse into skin following systemic administration of a drug. Principal of photochemical activation (Grotthuss–Draper Law) states that light must be absorbed by a chromophore for a chemical reaction to happen. Therefore, the first step in evaluation of photosafety of a chemical is to have an absorption spectrum conducted. If a substance absorbs light within the range of UV and visible light (290–700 nm), reactive species can be generated that can harm tissue. Additional testing is required to assess photosafety of a product that is intended for topical use, or a systemic medication that diffuses to photo-exposed skin.[1,2] This chapter discusses validated testing methods for the assessment of phototoxic properties of chemicals.

GENERAL CONSIDERATIONS FOR PHOTOIRRITATION TESTING

Skin is optically heterogeneous and via reflection, refraction, scattering, and absorption can modify the amount of radiation reaching deeper structures.[3] These protective properties can be affected by the topical application of different products. Vehicles can decrease the amount of light reflected, scattered, or absorbed.[4] They can also affect percutaneous absorption into the skin.[4,5] Physical and chemical properties of vehicles can affect absorption and photoproperties of applied medications; therefore, testing of vehicles along with the medications is required by regulating agencies.[1]

Approach to initial evaluation of phototoxicologic properties of a substance is summarized in Figure 3.4.

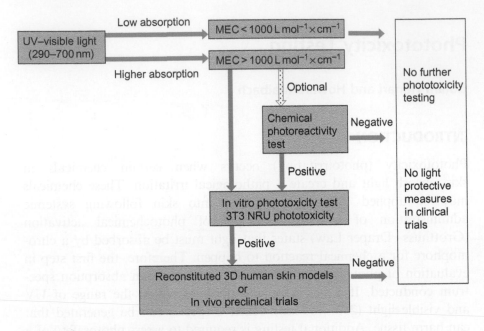

Figure 3.4 Outline of phototoxicity assessment strategies.

Predictive *In Vitro* Testing
Screening for UV and Visible Light Absorption
The initial assessment of photosafety of substances starts by their ability to absorb UV/visible light (wavelengths between 290 and 700 nm). Ability of chemicals to absorb light is an intrinsic property of that specific chemical. To measure how strongly a chemical absorbs light in a given wavelength, molar extinction coefficient (MEC) also called molar absorptivity is used. MEC is a constant for any given molecule under a specific set of conditions (e.g., solvent, temperature, wavelength) and reflects the efficiency with which a molecule can absorb a photon (typically expressed as $L \, mol^{-1} \times cm^{-1}$).[6]

Photoreactivity Testing Using Chemical Assays
Phototoxicity refers to tissue inflammation caused by interaction of light and a photoreactive chemical. A photoreactive chemical, at a sufficient tissue concentration, upon exposure to light can generate reactive oxygen species (ROS), singlet oxygen, and/or superoxide. Therefore, ROS generation following visible light or UV exposure can indicate potential phototoxicity of a chemical.

ROS assay is designed considering the above principle.[7,8] This assay has high sensitivity and low specificity, creating many false positives; therefore, a positive results at any concentration would only point out the need for further assessment. On the other hand, negative result on this assay, provided that the concentration of the test chemical is at least 200 μM, indicates a very low probability of phototoxicity and no further testing is required.[6]

Phototoxicity Tests Using In Vitro Assays
In Vitro 3T3 NRU Phototoxicity Test

In this assay, monolayer cell cultures of immortalized mouse fibroblast cell line Balb/c 3T3 are used to assess the cytotoxicity of a chemical in presence or absence of a noncytotoxic dose of simulated solar light. Cytotoxicity is expressed as a concentration-dependent uptake of neutral red (NR) dye (3-amino-7-dimethylamino-2-methylphenazine hydrochloride) 24 h after treatment with the test chemical and irradiation.[2] NR is a weak cationic dye that penetrates cell membranes by nondiffusion and accumulates intracellularly in lysosomes. In damaged cells, alteration of the lysosomal membranes and their fragility by a phototoxic reaction leads to decreased uptake and binding of NR. On this bases viable, damaged or dead cells can be distinguished. Concentration of test materials resulting in 50% reduction in NR uptake (means that cell viability is reduced by 50%), is called IC_{50} and is used to evaluate results.

Protocol follows the following steps (full details available on test guidelines 432, 2004).[2]

- A permanent mouse fibroblast cell line, Balb/c 3T3 is used.
- To test one chemical two sets of 96 well plates will be seeded with 1×10^4 cells per well and incubated for 24 h (37 C, 5% CO_2 in air). Both sets will go through the entire test procedure under identical conditions except that one plate is irradiated and one plate is kept in dark.
- On second day, test chemicals are added to wells. Test chemicals are dissolved in a physiologically buffered solution, free of protein and light absorbing components. Dilution series will be added to wells.
- Highest concentration should be physiologic, not exceeding 1000 μg/mL and osmolality should not exceed 10 mmolar. A geometric dilution series of eight test substance concentrations including a negative control or solvent are added to wells for 1 h (37 C, 5% CO_2 in air).

A known phototoxic chemical (Chlorpromazine) is added to each well as a positive control.

- One set of wells is exposed to $5 \, J/cm^2$ in the UVA range for 50 min, while the other set is kept in dark.
- All wells are subsequently rinsed with phosphate buffered saline and incubated in culture medium over night (37 C, 5% CO_2 in air).
- On third day NR in medium is added to each well (100 µL of 50 µg/mL solution) for 3 h.
- Wells are subsequently washed and optical density of the NR extract at 540 nm is measured in a spectrophotometer.
- Data will be evaluated by calculating, a photoirritation factor (PIF) or mean photoeffect (MPE).
- Interpretation of data is based on the above values as well as comparison of results between photoexposed and photopertected sets.

PIF is calculated based on the results of the NR uptake by comparing the IC_{50} with and without irradiation (PIF = IC_{50} without irradiation/IC_{50} with irradiation). MPE may also be calculated for interpretation of results.[2,9]

This test does not assess the phototoxic potency of a chemical and is not designed to evaluate photogenotoxicity, photoallergy, or photocarcinogenicity of a substance.

The 3T3 NRU phototoxicity test (3T3-NRU-PT) assay gained regulatory acceptance in all EU Member States in 2000, and in the Organization for Economic Co-operation and Development (OECD) Member States in 2004 as test guideline 432. Based on the original OECD test guidelines (test guidelines 432, 2004)[2] chemicals with MEC of $10 \, L \, mol^{-1} \times cm^{-1}$ were considered unlikely to be photoreactive and additional phototoxicity testing was not required for such products. Later on, the threshold was increased to $1000 \, L \, mol^{-1} \times cm^{-1}$.[6,10,11] In Step 4 version of International Committee on Harmonization (ICH) Expert Working Group in 2013, a reduction of the maximum test concentration from 1000 to 100 µg/mL is suggested.[12] Further in vitro testing is required for substances with higher MEC.

Test guideline 432 is widely used in the chemical and cosmetics industries and is considered a reliable predictive of acute phototoxicity effects in animals and humans in vivo.[2]

Disadvantages of this assay include:

- Only substances soluble in water can be tested.
- In vivo and in vitro results don't always correlate since bioavailability and biokinetics cannot be modeled.
- This assay can only use irradiation in the UVA and visible range; UVB wavelengths are excluded considering their cytotoxicity to Balb/c 3T3 cells. Considering that majority of phototoxic reactions involve UVA, this limitation is accepted.

Reconstituted 3D Human Skin Models

RhE models are 3D biostructures composed of cultured normal human keratinocytes, which form a multilayered epidermis including stratum corneum. Assessments of skin phototoxicity via RhE models are based on the premise that a phototoxic chemicals after exposure to light will damage cells and can be identified by their ability to lower cell viability. Then cell viability in RhE models is measured by enzymatic conversion of the vital dye MTT [3-(4,5-dimethylthiazol-2-yl)-2,5-diphenyltetrazolium bromide, thiazolyl blue] into a blue formazan salt that is quantitatively measured after extraction from tissues. Few 3D models have been used for in vitro phototoxicity testing including EpiDerm™, EpiSkin™, SkinEthic™.[13] EpiDerm underwent prevalidation studies founded by European Centre for the Validation of Alternative Methods (ECVAM) in 1999 and was found to generate good results. However, these models are costly, not appropriate for screening and may not be able to accurately predict safe, nonphototoxic concentrations of tested materials.[14–16]

A recommended protocol is outlined here[16]:

- 3D RhE models are used in six well plates; two sets of wells.
- Test material, positive or negative control, is applied directly on the tissue surface of the cultured tissue and incubated overnight (20 μL of the test material, or an adjusted dose based on the properties of the test material).
- On second day, the test material is rinsed off using buffered saline solution and one set of wells is irradiated using UVA solar simulator for 60 min/delivering a total of 6 J/cm^2.
- After irradiation, tissues are rinsed with saline and incubated again over night in fresh medium.

- On third day, cultures are transferred to 24 well plates containing 0.3 mL of MTT (1 mg/mL solution) and incubated for 3 h.
- Subsequently tissues are rinsed and transferred to an isopropanol solution. Isopropanol helps extract formazan from the tissues leading to the deep blue/purple color.
- Extracts are transferred to 96 well plates and absorption at 570 nm is measured in a spectrophotometer.
- Results are calculated based on the percentage of MTT reduction compared to control treated cultures.
- Phototoxic potential is predicted by >30% increase in cell toxicity in photoexposed cultured tissues.

These models have a structure similar to in vivo epidermis and contain both viable primary skin cells and skin barrier. Variety of chemicals, including complex mixtures, can be tested on 3D skin models. Barrier function of the stratum corneum can lead to more relevant results and less false positive reactions compared to highly sensitive monolayer cells. Higher concentration of tested materials (compared to monolayer cells) can be used, therefore more relevant to in vivo usage.[16] Despite the advantages of 3D models over the monolayer cells used in 3T3-NRU-PT model, theses models are more permeable compared to human skin and lack the dermal component.

Predictive *In Vivo* Testing

In vivo testing has been carried out in multiple lab animals, including rabbits, mice, guinea pigs, and swine. However, animal tests may create results not correlating to human testing. Maibach et al. proposed human phototoxicity testing[17] as well as Kaidbey and Kligman.[18] Human testing with approval of institutional review boards and after obtaining informed consent can be performed. Since theoretically almost any individual can develop phototoxic reaction, given sufficient exposure, these tests can be conclusive in a small number of volunteers. These reactions usually resolve quickly after removal of stimuli.[19]

Phototoxicity testing in human

- 10 consented healthy adults are enrolled.
- Two sets of test products and relevant controls (typically the vehicle) are applied on each side of the back skin, under occlusion for 6 h. Commercially available patch test chambers about 12 mm in

diameter are used. Additional control is done by application of one empty chamber on each side.

- Patches are removed after 6 h and only one side is radiated with UVA 20 J/cm². A solar simulator or a xenon arc lamp can be used as a source of radiation. Appropriate filtration with a Schott WG 345 filter blocks the erythemogenic UVC and UVB range. Longer wavelengths such as visible light and IR are filtered using coated dichromic mirror, water filter, and UG11 filters.
- Skin assessment is done immediately after radiation and at 24 and 48 h intervals. Reactions are graded and compared between radiated and unradiated sides. Reactions are graded from 0 to 5 (0 = no reaction, 1 = mild erythema ± scaling, 2 = moderate to strong erythema, 3 = moderate to strong erythema with a papular response, 4 = 3 + edema, 5 = vesicular eruption).

Reactions that are more prominent on the irradiated side indicate phototoxicity, while reactions that are equal on both sides refer to a nonphototoxic irritation.[19]

REFERENCES

1. Guidance for industry photosafety testing. <http://www.fda.gov/downloads/Drugs/.../Guidances/ucm079252.pdf> Accessed 01.08.14.

2. OECD. *Test no. 432: in vitro 3T3 NRU phototoxicity test*. Organisation for Economic Co-operation and Development; 2004. 10.1787/9789264071162-en.

3. Kornhauser A, Wamer WG, Lambert LA. Cellular and molecular events following ultraviolet irradiation of skin. In: Marzulli FN, Maibach HI, eds. *Dermatotoxicology*. 5th ed. Washington, DC: Taylor and Francis; 1996:189–230.

4. Anderson RR, Parrish JA. The optics of human skin. *J Invest Dermatol*. 1981;77:13–19.

5. Marzulli FN, Maibach HI. Photoirritation (phototoxicity, phototoxic dermatitis). In: Marzulli FN, Maibach HI, eds. *Dermatotoxicology*. 5th ed. Washington, DC: Taylor and Francis; 1996:231–237.

6. *ICH*. Guidance on photosafety evaluation of pharmaceuticals S10—Step 2 version—13 November, 2012. <http://www.fda.gov/downloads/Drugs/GuidanceComplianceRegulatory Information/Guidances/UCM337572.pdf> Accessed 15.01.14.

7. Onoue S, Hosoi K, Wakuri S, et al. Establishment and intra-/inter-laboratory validation of a standard protocol of reactive oxygen species assay for chemical photosafety evaluation. *J Appl Toxicol*. 2013;33(11):1241–1250.

8. Onoue S, Kawamura K, Igarashi N, et al. Reactive oxygen species assay-based risk assessment of drug-induced phototoxicity: classification criteria and application to drug candidates. *J Pharm Biomed Anal*. 2008;47(4–5):967–972.

9. Spielmann H, Balls M, Dupuis J, et al. The international EU/COLIPA in vitro phototoxicity validation study: results of phase II (blind trial). Part 1: the 3T3 NRU phototoxicity test. *Toxicol In Vitro*. 1998;12(3):305–327.

10. Henry B, Foti C, Alsante K. Can light absorption and photostability data be used to assess the photosafety risks in patients for a new drug molecule? *J Photochem Photobiol B*. 2009;96 (1):57−62.

11. *ICH*. Final concept paper S10: Photosafety evaluation of pharmaceuticals dated 8 April 2010. <http://www.ich.org/fileadmin/Public_Web_Site/ICH_Products/Guidelines/Safety/S10/Concept_Paper/S10_Final_Concept_Paper_June_2010x.pdf> Accessed March 2014, 2010.

12. Ceridono M, Tellner P, Bauer D, et al. The 3T3 neutral red uptake phototoxicity test: practical experience and implications for phototoxicity testing—the report of an ECVAM-EFPIA workshop. *Regul Toxicol Pharmacol*. 2012;63(3):480−488.

13. Netzlaff F, Lehr CM, Wertz PW, Schaefer UF. The human epidermis models EpiSkin, SkinEthic and EpiDerm: an evaluation of morphology and their suitability for testing phototoxicity, irritancy, corrosivity, and substance transport. *Eur J Pharm Biopharm*. 2005;60 (2):167−178.

14. Kejlova K, Jirova D, Bendova H, et al. Phototoxicity of bergamot oil assessed by in vitro techniques in combination with human patch tests. *Toxicol In Vitro*. 2007;21(7):1298−1303.

15. Liebsch M, Traue D, Barrabas C, et al. Prevalidation of the EpiDerm phototoxicity test. In: Clark D, Lisansky S, Macmillan R, eds. *Alternatives to Animal Testing II: Proceedings of the Second International Scientific Conference Organised by the European Cosmetic Industry*. Brussels/Newbury: CPL press; 1999:160−166.

16. Jones P. In vitro phototoxicity assays. In: Chilcott RP, Price S, eds. *Skin Toxicology*. England: John Wiley & Sons; 2008:169−183.

17. Maibach HI, Sams WM, Epstein JH. Screening for drug toxicity by wavelengths greater than 3100A. *Arch Dermatol*. 1967;95:12−15.

18. Kaidbey KH, Kligman AM. Identification of topical photosensitizing agents in humans. *J Invest Dermatol*. 1978;70(3):149−151.

19. Pearse AD, Anstey A. Clinical aspects of phototoxicity. In: Chilcott RP, Price S, eds. *Skin Toxicology*. England: Wiley; 2008:245−257.

CHAPTER 4

Allergic Contact Dermatitis: Clinical Aspects

Jean-Marie Lachapelle

INTRODUCTION

Two main advances in the field of dermato-immunology have changed the landscape of irritant contact dermatitis (ICD) and allergic contact dermatitis (ACD). First of all, ICD (innate immunity) and ACD (adaptive immunity) follow a similar pathway when a chemical xenobiotic penetrates into the skin tissues, after a disruption of the stratum corneum barrier, leading to an inflammatory response. The process remains limited in ICD, whereas, in ACD, the xenobiotic acts as an antigen (hapten) after triggering antigen-specific T lymphocytes, and this provokes the start of an allergic reaction.[1] This innovative concept explains why the clinical discrimination between ICD and ACD is often difficult (see later).

The second advance is the discovery of filaggrin (FLG) mutations, mainly encountered in atopic dermatitis (AD). It is well proved that FLG mutation carriers with self-reported dermatitis have an increased risk of contact sensitization to various allergens, whereas FLG mutations alone may not, or may only slightly, increase the risk of sensitization and, therefore, of ACD.[2] ACD is observed in daily life by the practicing dermatologist. In the vast majority of cases, its clinical presentation is an eczematous reaction. ACD is therefore synonymous with allergic contact eczema. But its clinical facets are so diversified that they need to be described, step by step, in separate sections.

The most comprehensive way to proceed methodically is to refer to the concept of the allergic contact dermatitis syndrome (ACDS).

ACD SYNDROME

We have developed the concept of "ACD syndrome" (ACDS).[3] A syndrome can be defined as a group of signs and symptoms that actively indicate or characterize a disease.[4]

Applied Dermatotoxicology. DOI: http://dx.doi.org/10.1016/B978-0-12-420130-9.00004-9

A similar approach was made previously regarding irritation, i.e., the ICD syndrome[5] and contact urticaria, i.e., the contact urticaria syndrome.[6] The concept of ACDS considers the various facets of contact allergy, including morphological aspects and staging by symptomatology.

The three stages of ACDS can be defined as follows:

1. Stage 1. Skin signs and symptoms are limited to the site(s) of application of contact allergen(s).
2. Stage 2. There is a regional dissemination of signs and symptoms (via lymphatic vessels) extending from the site of application of allergen(s).
3. Stage 3. Corresponds to the hematogenous dissemination of either ACD at a distance (stage 3A) or systemic reactivation of ACD (stage 3B).

STAGE 1 OF ACDS

Morphological Aspects

As emphasized previously, the clinical picture of ACD is eczematous in most cases, but it varies depending on its location and duration. In most instances, acute eruptions are characterized by erythema and papules, vesicles (often coalescent), or bullae, depending on the intensity of the allergic response (Figure 4.1). In severe cases, this can lead to abundant oozing, followed some time later by the occurrence of yellowish crusts.

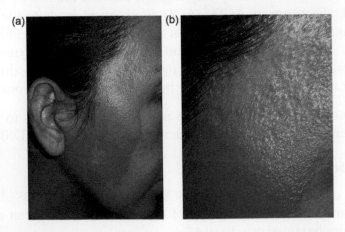

Figure 4.1 (a) Acute ACD to a hair dye, extending to the cheeks. (b) A closer view clearly illustrates the presence of aggregated vesicles. The paraphenylenediamine patch test was positive.

In milder cases, erythema and tiny papules are the only symptoms of acute ACD. When it is a first contact with the culprit allergen, the eruption usually appears after 5–7 days. After repeated contact with the same (or chemically closely related) allergen, the first symptoms appear after 1–3 days. They subside in a few days, when the contact is interrupted. It is noteworthy that, whatever their degree of severity, the margins of the lesions are most often ill-defined, extending beyond the site of application of the allergen(s). This is in contrast with the lesions of ICD, which are usually sharply demarcated.

Some areas deserve special attention, exhibiting particular morphological variants.

For instance, acute ACD of the scalp is not only erythematous, but also predominantly scaly. Edema is conspicuous on the eyelids, scrotum, penis, and, more discretely, on the lower lobe of the ear. This particularity is linked with the extreme laxity of these tissues. Allergic contact stomatitis or vulvitis is diffusely erythematous, sometimes edematous, without vesiculation.

Chronic ACD of nearly all cutaneous sites is far more monomorphous; it presents as a thickened scaling, occasionally fissured dermatitis, with or without accompanying vesiculation (Figure 4.2).

Pruritus is always present, very often severe. But, of course, it is by no means specific for ACD; all varieties of eczema are indeed pruritic, as, for instance, AD.

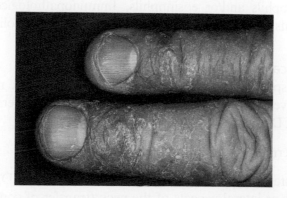

Figure 4.2 Chronic ACD of the fingers. Lesions are erythematous, scaly, and discretely fissured. The epoxy resin patch test was positive.

The histopathological picture of ACD is a typical example (sometimes called the "prototype" of a spongiotic dermatitis). Its characteristics have been described in full detail,[7] but it does not offer any clue to the diagnosis. All types of eczematous diseases are indeed spongiotic, and the differences can be considered minor and subtle.

If the vast majority of cases of ACD are eczematous, other morphological variants are infrequently observed. They are manifold and can be described as follows:

Purpuric ACD. This variant is mainly observed on the lower legs and/or feet and has been reported with a variety of allergens (i.e., anti-inflammatory nonsteroidal topical drugs, textile dyes, etc.). Purpuric lesions are prominent or associated with eczematous symptoms (sometimes bullous on the lower part of legs and/or feet). They may occur in other regions. Purpura is the clinical manifestation of the extravasation of erythrocytes into dermal tissue and epidermis.

Lichenoid ACD. Lichenoid ACD is rare. Its clinical features mimic lichen planus (e.g., from metallic dyes in tattoos or from corals). Oral lichenoid ACD looks like oral lichen planus (e.g., from dental amalgams).

Pigmented ACD. It is mainly reported in Oriental populations.

Lymphomatoid ACD. This variant cannot be defined as a clinical distinctive entity; it is based only on histopathological criteria. Clinical signs (nondiagnostic) are erythematoedematous plaques, sometimes very infiltrated, at the site(s) of application of contact allergen(s). Histopathological examination reveals the presence of an important dermal (and sometimes subdermal) infiltrate, displaying features of pseudolymphoma, i.e., mainly lymphohistiocytic with a few neutrophils and/or eosinophils. Immunopathological investigation permits the exclusion of malignant lymphocytic proliferation.

Topographical Variants

ACD can display some topographical peculiarities that may be misleading for every trained dermatologist. This mainly refers to cases of "ectopic" ACD and airborne ACD.

Ectopic dermatitis can follow these:

1. Autotransfer. A typical example is nail lacquer ACD, located on the eyelids or lateral aspects of the neck (transfer of contact allergen by fingers).

2. Heterotransfer. The often-quoted example is transfer of the allergen(s) to the partner. Such events have been described as connubial ACD, consort ACD, ACD *per procurationem*, or ACD « by proxy »; note that in these circumstances, the patient applying the allergen is usually free of any symptoms. To be acquainted with this particular entity, it is worth reading a recent review, illustrated by many examples.[8]

Airborne ACD is another pitfall for clinicians.

Allergen(s) is(are) transported by air as dust particles, vapors, or gasses. In most cases, ACD involves the face, neck, and/or décolleté. There is usually no spared area, contrary to phototoxic and/or photo-allergic contact dermatitis (see later). Limits of eczematous lesions are ill-defined. There is no definite clue to make a clinical distinction between irritant and allergic airborne contact dermatitis. Patch testing is therefore of utmost diagnostic value. The occurrence of airborne ACD and airborne ICD is underestimated because reports omit the term "airborne" in relation to dust or volatile irritants and/or allergens. An updated list of references is available.[9]

The main sources of airborne ACD are occupational allergens,[10] cosmetics, and plants.

STAGE 2 OF ACDS

Stage 2 of ACDS is linked with the regional dissemination via lymphatic vessels of ACD from the primary site of application of the allergen(s). In most cases, ACD lesions are more pronounced at the site(s) of application of the allergen(s), and disseminating lesions fade progressively from the primary site. They appear as erythematous or erythematovesicular plaques with poorly defined margins. In some other cases, extending lesions are more pronounced than those located at the primary site. This paradoxical observation is not fully understood. It is hypothesized that, by dissemination, allergens are more concentrated on these areas than at the primary site of application. It sometimes occurs with, for example, nonsteroidal anti-inflammatory drugs or antibiotics.

Three clinical variants of regional dissemination involve more intricate immunological mechanisms. These include:

a. True erythema multiforme lesions, displaying both clinical and histopathological signs of erythema multiforme. Such reactions have been reported with several allergens.[3] The most frequently quoted are woods and plants (*Dalbergia nigra*, pao ferro, *Primula obconica*, etc.), metals (nickel, cobalt), paraphenylenediamine, and epoxy resin.

b. Erythema multiforme—like lesions presenting clinically as "targeted" lesions typical of erythema multiforme, but histopathological signs of a spongiotic dermatitis, characteristic for eczematous dermatitis.[3]

c. The two syndromes (a) and (b) are well documented in some publications, whereas in some others there is no clear-cut distinction between both groups due to a lack of histopathological investigations.

It is still difficult to understand the pathomechanisms involved in all these "noneczematous reactions." But, at the present stage of knowledge, the interaction between CD8+ T cells and CD4+ T cells is modified in some way. Further studies are needed to clarify these particular events.

In the meantime, widespread secondary lesions can occur simultaneously at a distance of the primary site (stage 3A).

STAGE 3 OF ACDS

Stage 3 of ACDS includes two distinct entities, leading sometimes to unexpected confusion in the current literature. A clear-cut distinction between both entities is detailed in the following sections.

Stage 3A of ACDS

Stage 3A of ACDS can be defined as a generalized dissemination of skin lesions—via blood vessels—from the primary site of application of the allergen. It is considered that the allergen penetrates through normal and/or lesional skin and reaches distant skin sites (hematogenous dissemination) where it provokes secondary (or "ide") reactions. These reactions appear as symmetrical erythematous, sometimes slightly elevated plaques, more rarely vesicular or squamous. They are of "pompholyx type" on palmar and/or plantar skin (see later).

Malten et al.[11] coined the term "chemides" to describe the various skin manifestations at distant sites. Chemides are always concomitant with ACD lesions at the primary site(s) of application of the allergen.

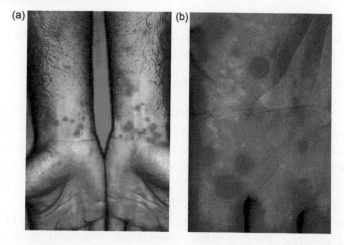

Figure 4.3 Stage 3A of ACDS. True erythema multiforme symmetrical lesions at distant sites (hematogenous dissemination) from the primary site of sensitization (ides). Contact allergy to dalbergiones (Dalbergia nigra).

Sugai added to Malten's initial description some clinical variants, such as true erythema multiforme lesions (Figure 4.3) and erythema multiforme–like lesions[12]; these types of lesions being similar to those reported in stage 2 of ACDS.

Stages 2 and 3A of ACDS can be present simultaneously in the same individual.

The concomitant occurrence of both stages of lesions illustrates the clinical complexity of ACDS.

Among contact allergens involved in stage 3A of ACDS and reported in the literature, some deserve special interest: paraphenylenediamine, cobalt, nickel, mercury, mercuric chloride, corticosteroids, and nonsteroidal anti-inflammatory agents.

Stage 3B of ACDS
Stage 3B of ACDS has been described as follows:

1. Baboon syndrome.[13] This term is not satisfactory since it tends erroneously to circumscribe symptoms to limited skin areas, i.e., buttocks, groin, and perineal region; therefore it does not take into account other skin sites which are involved as well.
2. Fisher's systemic contact dermatitis or, more precisely systemic contact-type dermatitis.[14]

In essence, the most appropriate expression could be systemic reactivation of allergic contact dermatitis (SRCD).[3] It considers the chain of events resulting in the occurrence of stage 3B of ACDS.

Two successive episodes are involved in this event:

1. First episode: A first event of ACD to a well-defined contact allergen (allergen 1) has occurred in the past (weeks or even years before episode 2). All clinical symptoms have vanished completely when contact with allergen 1 has ceased. Sometimes, patients have forgotten about it; this emphasizes the need for a complete clinical history (a general rule in the field of contact allergy).
2. Second episode: In some cases, the substance (molecule 1) is introduced systemically (ingestion, inhalation, injection), and its use is followed by a more or less generalized skin rash, usually in a symmetrical pattern (as in stage 3A of ACDS). The molecule is the true allergen (allergen 1). In other cases, another substance (molecule 2) is used systemically and provokes SRCD.

This could be related with two different mechanisms:

a. Molecule 2 is chemically closely related to molecule 1. Both are allergenic, and there is cross-sensitization. Molecule 2 is therefore considered allergen 2.
b. Another possibility is that molecules 1 and 2 are not allergenic as such, but both are transformed into another common molecule, which is the allergen (responsible for episodes 1 and 2).

The clinical signs observed in stage 3B of ACDS share a similar pattern with skin lesions observed in stage 3A of ACDS. The only difference is that in stage 3B, no current skin contact does occur (episode 2).

ICD VERSUS ACD: CLINICAL ASPECTS

The current view is that differential diagnosis between ICD and ACD, based on clinical grounds, is difficult. Some criteria have been advocated in the past, but most of them suffer from many exceptions.

Topography
Acute ICD (erythematous and sometimes vesicular and/or bullous) appears rapidly, usually minutes to few hours after exposure. Chronic

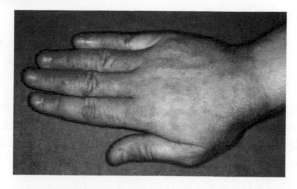

Figure 4.4 ICD. Pruritic, discretely painful, sharply demarcated plaque of the dorsum of the hand, due to repeated contact with household detergents.

ICD is characterized by hyperkeratosis, fissuring, glazed, or scaled appearance of the skin; but it is often claimed that, in both cases, lesions are characteristically sharply circumscribed to the contact area (Figure 4.4). But it is not true in all cases; sometimes dermatitis may be generalized, depending on the nature of the exposure, similarly to ACD. As mentioned previously, the limits of ACD are most often ill-defined. Dissemination of the dermatitis with distant lesions may occur, but in some cases, eczematous lesions are strictly limited to the site of contact, for instance nickel dermatitis under a watch or a jeans' button.

Timing of Evolution

When the contact with the chemical has ceased, lesions of ICD start to heal rather quickly (the so-called decrescendo phenomenon), whereas lesions of ACD may still worsen (the so-called crescendo phenomenon), but, here again, it is not always the case. Therefore, this presumed criterion of differential diagnosis is, in reality, breached.

Subjective Symptoms

Pruritus is the main symptom of ACD, whereas symptoms of ICD are often burning, stinging, pain, and soreness of the skin, but, not infrequently, pruritus is also present. By the way, in general dermatology, pruritus is considered an unreliable symptom.

Conclusion

Differential diagnosis between ICD and ACD, based exclusively on clinical grounds, is never conclusive.

HAND ECZEMA (HAND DERMATITIS) AND ACD

Hand eczema deserves special attention, due to the very unique and multifacetted anatomoclinical characteristics of the hand, and to its repeated specific environmental exposure.

The problems raised by hand eczema are therefore complex; there are many varieties of eczematous conditions not at all related to ACD, which are fully described in a recent textbook.[3] The reader is invited to refer to it, for a better understanding of this complexity.

Focusing on ACD, there are three core messages.

Various Clinical Aspects of ACD of the Hands

Various clinical aspects have to be taken into consideration.

On the backs of the hands, ACD (Figure 4.5) can be either acute (erythema, vesicles, bullae, and sometimes oozing) or chronic (erythematous and scaly lesions), usually very pruriginous (Figure 4.5). Exceptionally, lesions are round-shaped, mimicking nummular dermatitis, and this morphological particular aspect is puzzling, and, in fact, remains obscure.

On the palms, a similar scenario does occur; either acute reactions of the "pompholyx" type, or chronic (palmar erythematosquamous dermatitis).

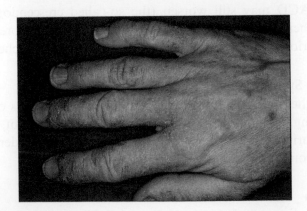

Figure 4.5 Chronic very typical ACD of the dorsum of the hands in a bricklayer. The potassium dichromate patch test was positive.

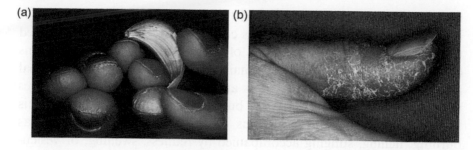

Figure 4.6 Fingertip dermatitis. (a) ACD to garlic in a female cook. The diallyldisulfide (allergen of garlic) patch test was positive. (b) PCD to monkfish in a fishmonger. The prick test to monkfish was positive (reading at 30 min).

Fingertip dermatitis is another occasional picture of ACD (Figure 4.6a). Chapping of the fingertips is common. Painful crevices and bleeding occur in severe cases. We have stressed that fingertip dermatitis limited to the thumb and index (and eventually medius) of one or both hands frequently implies irritant (frictional and/or chemical) or allergenic factors. In those cases, fingertip dermatitis may be typical of (a) ICD, (b) ACD, or (c) protein contact dermatitis (see later). We have coined the term "gripping form" of fingertip dermatitis.[3] Such considerations are far too simple; in many of these cases, the skin condition remains unclear, and it is therefore considered endogenous, and environmental factors playing only an adverse role. When some fingers are randomly involved, whereas others are spared, or in case of complete involvement of all fingers of both hands, etiology is even more obscure.

Overlapping of Eczematous Lesions of the Hands

Overlapping of eczematous lesions of the hands is frequently encountered in daily practice. The most concomitant occurrence of overlapping is ICD, ACD, and AD. Thus, this problem has to be evaluated and solved by the clinician. The classification of hand dermatitis is therefore difficult (either morphological or etiological), particularly in the field of occupational dermatology. It is the reason why, in different European countries, an inclusive term has been created: "hand eczema." This new concept (pros and contras) is worthwhile to be considered, because it takes into account the fact that patients can be affected along the years by different variants of eczema.[15,16]

Protein Contact Dermatitis

Protein contact dermatitis (PCD) is considered a specific entity, related to the penetration into the skin of animal and/or vegetal proteins (MW > 1000) and enzymes. It is mainly observed during occupational activities (Figure 4.6b). The hands are particularly involved. It shares many clinical aspects with ACD, but a distinctive feature of PCD is the fact that the patient complains of immediate symptoms, such as burning, itching, stinging accompanied by redness, swelling, or vesiculation when handling the allergen. It appears that PCD could appear more frequently in patients suffering from AD than in nonatopics.

Clinical variants do exist:

- Fingertip dermatitis. Mainly but not exclusively of the "gripping type."
- Chronic paronychia. This is a common variant mainly observed in patients who have chronically wet hands. Wet foods are a combined source of factors, where the food may be an irritant or an allergic contactant.[17]

PHOTOALLERGIC CONTACT DERMATITIS

Photoallergic contact dermatitis (PACD) is produced when sensitization occurs from the combination of skin contact with a compound together with ultra violet light (UVL, generally UVA) exposure. In these cases, the hapten (allergen) requires UVL to be fully activated. Globally, the clinical symptoms are similar to those of ACD, but limited to light-exposed sites, i.e., the face, neck (Figure 4.7), dorsal

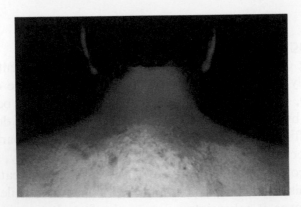

Figure 4.7 PACD to a sunscreen. Erythematous and vesicular ill-defined lesions. Covered sites are spared. The octocrylene patch test was positive.

hands, and forearms. It spares shaded sites such as the upper eyelids, submental area, and postauricular areas.

PACD versus airborne (nonphotoallergic) ACD is an important matter of differential diagnosis. In the latter, spare areas of PACD are usually involved, which is of great help.

Another potential pitfall of diagnosis is to confound PACD with polymorphic light eruption (PLE, chronic actinic dermatitis, actinic reticuloid) an idiopathic, severe, chronic photodermatosis, which occurs most often in men, middle-aged, or older and characterized by infiltrated, erythematous, shiny papules on an eczematous background on exposed areas, often with involvement of covered sites. The patients react to UVA, UVB, and visible light. However, when the history and the physical examination suggest the possibility of PACD, it has to be kept in mind that PACD can in fact be superimposed to PLE.

ACD: THE PATCH TEST, FIRST-LINE DIAGNOSTIC TOOL

The patch test remains the "gold standard" in the diagnosis of ACD. It is out of the scope of this review devoted to clinical aspects of ACD, to describe its technical aspects in detail, that are fully explained in classical textbooks of dermato-allergology. It has to be stressed that patch test results have to be scrutinized very cautiously, more than in the past, due a better knowledge of mechanisms involved in ACD, as explained in the introductory note.

Nevertheless, several items deserve special attention.

Reading Time

It is universally codified. The best approach is reading at 48 h, 96 h, and 7 days (for late reactions characteristic of some allergens and/or for late reactors). If only one reading is available, 72 h is the best choice.

Scoring Patch Test Results

The International Contact Dermatitis Group (ICDRG) scoring system[18] (Table 4.1) remains unequivocal; other proposals of classification are still under debate.

Table 4.1 Scoring of Patch Test Reactions	
Score	Interpretation
–	Negative reaction
? +	Doubtful reaction; faint erythema only
+	Weak (nonvesicular) reaction; erythema, slight infiltration
+ +	Strong (edematous or vesicular) reaction; erythema, infiltration, vesicles
+ + +	Extreme (bullous or ulcerative)
IR	Irritant reactions of different types
NT	Not tested

Dealing with Doubtful Reactions (? +)

? reactions (questionable faint or macular nonpalpable erythema cannot be interpreted as a proven allergic reaction. Repeating patch testing and/or using additional tools of diagnosis is mandatory (see later).

Clinical Relevance of Patch Test Reactions

Reading patch test reaction cannot be limited to scoring as positive or negative. Scoring in itself has no meaning if it is not linked in some way with the medical history of the patient. In other words, a positive patch test (and to some extent a negative patch test) has no interest, if it is not labeled as relevant or nonrelevant;

Determining past and/or current relevance or nonrelevance requires a very detailed investigation, including anamnestic data and additional testing procedures.[19]

Photopatchtest

Photopatchtesting (PPT), simply stated, is patch testing with the addition of UV radiation to induce formation of the photoallergen, and scoring criteria are the same as those described for plain patch testing (except they are labeled Ph + , Ph + +, Ph + + +).

PPT is intended to detect the responsible photoallergen(s) in PACD.

THE SERIES OF PATCH TEST ALLERGENS

There are several series of patch test allergens marketed by different companies (Brial®, Chemotechnique®, Trolab® Smart Practice®, etc.). In addition, there is a ready-to-use device, the True Test® Smart Practice, limited to some allergens.

The Baseline (Standard) Series

The baseline (standard) series is the core of any diagnostic investigation as regards ACD. Its contents vary between different parts of the world, in relationship with the predominance of local significant allergens. Therefore, flexibility is the rule, and adaptations are constantly needed, following environmental changes. Some allergens can become obsolete, whereas others are emerging.[20]

In most cases, the baseline series is the first step of investigation, i.e., an initial guide for the clinician, but is insufficient for the elucidation of the problem. Therefore, additional series have been built up along the years, orientated to each individual case; anamnestic data play an important role in the choice.

Additional Series of Patch Tests

The main series available nowadays are the following: bakery, corticosteroid, cosmetic, epoxy resin, hairdressing, isocyanate, metals, (meth) acrylate, plastics and glues, rubber additives, textile dyes, and finish. There are also series dedicated to frequent cases of ACD; for instance, shoe dermatitis or plant dermatitis.

THE PANOPLY OF ADDITIONAL DIAGNOSTIC TOOLS

As emphasized earlier, the patch test has its own limits in the diagnostic approach. In recent years, it has been completed by a series of complementary tests which are of great help for the clinician. They correspond to well-defined procedures of use. They are quoted in this review, and the reader is invited to consult the hereafter referred papers, which provide a fully detailed explanation of each of them.[21]

- The strip patch test[22]
- The open test
- The semi-open (or semi-occlusive) test[23]
- The Repeated Open Application Test (ROAT)[24] It has to be noted that the provocative use test (PUT) is synonymous with the ROAT.
- The spot tests.[21]

TESTING END-PRODUCTS WITH UNKNOWN COMPONENTS

Some materials (solid or liquid), brought by the patients (mainly, but not exclusively in occupational dermatology) are unknown in terms of ingredients.

The strategy is threefold:

a. Try to obtain information from the manufacturer, as complete as possible.
b. Test with the end-products as such when they are solid material and often diluted when liquids are concerned (patch tests, open tests, semi-open tests, etc.).
c. For solid materials, it is advised to use the ultrasound bath technology to extract potential allergens, which are tested at the proper concentration.[25]

TREATMENT

The only efficacious treatment is to eradicate the culprit allergen(s). Symptomatic treatments, mainly topical corticosteroids, can be of help to reduce the symptomatology, before a definitive diagnosis has been established.

REFERENCES

1. Nosbaum A, Nicolas J-F, Lachapelle J-M. Pathophysiology of allergic and irritant contact dermatitis. In: Lachapelle J-M, Maibach HI, eds. *Patch Testing and Prick Testing: A Practical Guide. Official Publication of the ICDRG*. 3rd ed. Berlin: Springer; 2012:3–9.

2. Thyssen JP, Linneberg A, Ross-Hansen K, et al. Filaggrin mutations are strongly associated with contact sensitization in individuals with dermatitis. *Contact Dermatitis*. 2013;68:273–276.

3. Lachapelle J-M. Allergic contact dermatitis syndrome. In: Lachapelle J-M, Maibach HI, eds. *Patch Testing and Prick Testing: A Practical Guide. Official Publication of the ICDRG*. 3rd ed. Berlin: Springer; 2012:14–23.

4. Grosshans E, Lachapelle J-M. Signs, symptoms or syndromes? *Ann Dermatol Venereol*. 2008;135:257–258.

5. van der Valk PGM, Maibach HI. *The Irritant Contact Dermatitis Syndrome*. Boca Raton, FL: CRC Press; 1995:393 pp.

6. Maibach HI, Johnson HL. Contact urticaria syndrome. Contact urticaria to diethyltoluamide (immediate-type hypersensitivity). *Arch Dermatol*. 1975;111:726–730.

7. Lachapelle JM, Marot L. Histopathological and immunohistopathological features of irritant contact dermatitis. In: Johansen JD, Frosch PJ, Lepoittevin JP, eds. *Contact Dermatitis*. 5th ed. Berlin: Springer; 2011:167–177.

8. Mc Fadden J. Proxy contact dermatitis or contact dermatitis "by proxy" (consort or connubial dermatitis). [chapter 10] In: Lachapelle J-M, Bruze M, Elsner PU, eds. *Patch Testing Tips. Recommendations from the ICDRG*. Berlin: Springer; 2014:115–122.

9. Swinnen I, Goossens A. An update of airborne contact dermatitis: 2007–2011. *Contact Dermatitis*. 2013;68:232–238.

10. Lachapelle J-M. Occupational airborne contact dermatitis: a realm for specific diagnostic procedures and tips. [chapter 9] In: Lachapelle J-M, Bruze M, Elsner PU, eds. *Patch Testing Tips. Recommendations from the ICDRG.* Berlin: Springer; 2014:101–114.

11. Malten KE, Nater JP, Van Ketel WG. *Patch Testing Guidelines.* Nijmegen: Dekker and van de Vegt; 1976:135 pp.

12. Sugai T. Contact dermatitis syndrome (CDS). *Environ Dermatol (Nagoya).* 2000;7:543–544.

13. Andersen KE, Hjorth N, Menné T. The baboon syndrome: systemically induced allergic contact dermatitis. *Contact Dermatitis.* 1984;10:97–101.

14. Rietschel RL, Fowler Jr JF. *Systemic contact-type dermatitis. Fisher's Contact Dermatitis.* 6th ed. Hamilton: BC Decker; 2008:110–124

15. Lerback A, Kyvik KO, Ravn H, Menné T, Agner T. Clinical characteristics and consequences of hand eczema—an 8 year follow-up study of a population-based twin cohort. *Contact Dermatitis.* 2008;58:210–216.

16. Veien NK, Hattel T, Laurberg G. Hand eczema: causes, course and prognosis. *Contact Dermatitis.* 2008;58:330–334.

17. Lachapelle J-M, Maibach HI. Protein contact dermatitis. In: Lachapelle J-M, Maibach HI, eds. *Patch Testing and Prick Testing: A Practical Guide. Official Publication of the ICDRG.* 3rd ed. Berlin: Springer; 2012:153–156.

18. Wilkinson DS, Fregert S, Magnusson B, et al. Terminology of contact dermatitis. *Acta Derm Venereol.* 1970;50:287.

19. Lachapelle J-M. A proposed relevance scoring system for positive allergic patch test reactions: practical implications and limitations. *Contact Dermatitis.* 1997;36:39–43.

20. Castelain M, Assier H, Baeck M, et al. The European Standard Series and its additions: are they of any use in 2013? *Eur J Dermatol.* 2014,24:15–22.

21. Lachapelle JM, Maibach HI. Additional testing procedures and spot tests. In: Lachapelle J-M, Maibach HI, eds. *Patch Testing and Prick Testing: A Practical Guide. Official Publication of the ICDRG.* 3rd ed. Berlin: Springer; 2012:113–128.

22. Dickel H, Kamphowe J, Geier J, et al. Strip patch test VS. conventional patch test: investigation of dose-dependent sensitivities in nickel- and chromium-sensitive subjects. *JEADV.* 2009;23:1–8.

23. Goossens A. Semi-open (or semi-occlusive) tests. [chapter 11] In: Lachapelle J-M, Bruze M, Elsner PU, eds. *Patch Testing Tips. Recommendations from the ICDRG.* Berlin: Springer; 2014:123–127.

24. Hannuksela M, Salo H. The repeated open application test (ROAT). *Contact Dermatitis.* 1986;14:221–227.

25. Bruze M. The use of ultrasonic bath extracts in the diagnostics of contact allergy and allergic contact dermatitis. [chapter 12] In: Lachapelle J-M, Bruze M, Elsner PU, eds. *Patch Testing Tips. Recommendations from the ICDRG.* Berlin: Springer; 2014:129–142.

Toxicology of Skin Sensitization

Golara Honari and Howard Maibach

INTRODUCTION

Allergic contact dermatitis syndrome is caused by allergen-specific T-cell-mediated complex processes. Skin sensitization occurs when a susceptible individual is exposed to a chemical with potential to sensitize. Reexposure to the same allergen in a sensitized person can mount a localized or widespread skin inflammation. Assessment of skin sensitization hazard of chemicals is an important component of the safety assessment of chemicals. In the absence of in vitro validated models, the predictive testing for skin sensitization potential currently has relied on animal testing. Skin sensitization testing, at least for cosmetic ingredients, has been banned in the Europe. Animal testing with completed cosmetic products in European Union was prohibited in 2004; testing ban for cosmetic ingredients was enforced in 2009, followed by a marketing ban as of March 2013.[1] These directives pose an urgent challenge for scientists in search of alternative methods.

Currently validated methods to predict sensitizing potentials are mostly animal based. However, increasing understanding of the key steps of sensitization process has led to the development of few in vitro predictive methods. Recently two in vitro assays have been validated (Direct Peptide Reactivity Assay (DPRA) and the KeratinoSens™) for classifying a substance as a skin sensitizer. Formal validation of these assays is required before they can fully replace in vivo assays for hazard identification.[2] Skin sensitization is an intrinsic property of a chemical substance and is referred to as a hazard; the likelihood that this hazard will be expressed is called risk. Risk is a function of hazard potency and exposure (Risk = Hazard × Exposure). The likelihood of a particular person developing sensitivity is the function of his or her individual susceptibility.[3]

This chapter reviews basic mechanisms of contact sensitivity, o. cially accepted animal models including the Mouse local lymph noc assay (LLNA) and its nonradioactive modifications (LLNA-DA and the LLNA-BrdU Elisa), the Guinea Pig Maximization Test, Buehler occluded patch test in the guinea pig, recently validated in vitro assays (DPRA and the KeratinoSens) and the human Cell Line Activation Test (h-CLAT) which is under validation by European Centre for the Validation of Alternative Methods (ECVAM).

PATHOGENESIS

Allergic contact dermatitis is a delayed-type hypersensitivity response composed of an induction (sensitization) phase and an effector (elicitation) phase.[4]

Induction has multiple steps:

- Penetration of allergen into the skin and binding to skin components. An allergen needs to have certain physicochemical properties to be able to induce sensitivity. Most contact allergens are small, molecules with a molecular weight less than 500 Da, and referred to as hapten. Larger molecules can hardly penetrate the stratum corneum and cannot act as allergens.[5] At the same time, haptens are too small to be antigenic; they need to form a hapten–protein complex to activate antigen-presenting cells. Concept of protein/peptide haptenation is crucial to form immunogenic structures and is focus of much research to develop predictive in vitro sensitization.[6,7] Allergen in skin bind major histocompatibility complex (MHC) proteins, which are encoded by histocompatibility antigen (HLA) genes and are present on epidermal antigen-presenting cells (Langerhans cells).[6,8]
- Next step is activation of Langerhans cells by haptens. Langerhans cells are responsible for internalization and processing of the haptens. Allergen induced production of inflammatory cytokines and multiprotein complexes, termed inflammasomes, induces migration of the haptenized Langerhans cells and dendritic cells into the T-cell-rich paracortical areas of regional lymph nodes.[8,9]
- In the paracortical areas of the lymph node, naive T cells will recognize allergen–MHC molecule complexes to transform into allergen-specific T cells.

Proliferation of allergen-specific T cells in the draining lymph node is supported by interleukin (IL)-1, released by the allergen-presenting cells and IL2 released by activated T cells. IL2 is a potent inducer of T-cell proliferation.[10]

- Systemic increate of effector-memory T cells. In the absence of further allergen contacts, their frequency in blood gradually decreases.

Effector phase starts upon reexposure to allergen, which leads to activation of effector-memory T cells, release of multiple cytokines and chemokines, leading more inflammatory response and clinical presentation of dermatitis.

It generally takes about 4 days to few weeks to become sensitized, but elicitation can happen within 1 day up to a week.

Predictive In Vitro Testing

Understanding key biological mechanisms of the induction phase helps designing in vitro test models. It is unlikely for a single alternative method to provide sufficient information to replace the in vivo models. Integrated data from different alternative testing and nontesting methods is required to address this endpoint.[11]

The key steps for adverse outcome pathway (AOP) for skin sensitization include:[11,12]

- Skin bioavailability: ability of a chemical to penetrate skin and reach the site of haptenation
- Haptenation: ability of a chemical to form covalent binding to skin proteins
- Release of proinflammatory signals by epidermal keratinocytes
- Activation and maturation of dendritic cells
- Migration of dendritic cells from skin to the regional lymph nodes and presentation of the antigen to T cells
- Proliferation of memory T cells (lymphocytes capable of being stimulated and activated specifically by the haptenated protein).

Two mechanistically relevant assays are considered by ECVAM.

Direct Peptide Reactivity Assay

The DPRA is an in chemico assay which addresses the process of haptenation. Haptenation is the covalent binding of haptens to skin proteins and is considered the molecular initiating event of the skin sensitization. Haptenation is assessed by measuring depletion of synthetic heptapeptides containing cysteine or lysine, after 24 h of incubation with the test substance. The DPRA test method underwent EU Reference Laboratory for Alternatives to Animal Testing (EURL ECVAM) coordinated validation studies, and recommendations on this test were released in November 2013.[13] The DPRA is designed for screening the sensitization potential of chemicals. Test chemicals are incubated for 24 h with synthetic heptapeptides containing either cysteine (peptide/chemical ratio in the reaction mixture 1:10) or lysine residues (peptide/chemical ratio in the reaction mixture 1:50). Peptide depletion is measured by high-performance liquid chromatography with UV detection (UV-HPLC).[14] Chemicals are classified into four reactivity categories based on average peptide depletion values for cysteine and lysine into minimal, low, moderate, and high reactivity.[15] Sensitizing potency is assumed to correlate with the reactivity class. The accuracy of the DPRA for distinguishing sensitizers from nonsensitizers in 133 chemicals in relation to local lymph node assay (LLNA) is reported to be 86% (87% sensitivity, 83% specificity).[13] A draft of the protocol by EOCD is available for more details.[16]

Keratinocyte-Based ARE-Nrf2 Luciferase Reporter Gene Test Method: KeratinoSens

This assay addresses the activation of a key pathway in keratinocytes in response to potential sensitizers. The antioxidant/electrophile response element (ARE)-dependent pathway in keratinocytes is the second key step in skin sensitization. Nuclear factor erythroid 2-related factor 2 (Nrf2) is a transcription factor with a key role in promoting the expression of genes coding for cyto-protective proteins, following an electrophilic or oxidative stress. The activity of Nrf2 is primarily regulated by the cysteine-rich Keap1 sensor protein (Kelch-like ECH associated protein 1). Although other pathways may be involved in this regulation, Keap1-Nrf2-ARE pathway is considered a major regulator of cyto-protective responses to oxidative and electrophilic stress. While majority of skin sensitizers are electrophiles reacting with nucleophilic centers in skin proteins, activation of this pathway is a relevant measure of skin sensitization potential of a chemical.[17,18]

KeratinoSens in vitro test method is developed based on the above mechanism. It has gone through EURL ECVAM coordinated idation studies, and recommendations on this test were released on ɔruary 2014.[17]

A recommended protocol is outlined here[18]:

mmortalized adherent cell line derived from HaCaT human keratiɲocytes transfected with a selectable plasmid (plasmid contains a luciferase gene under the transcriptional control of ARE). This allows quantitative measurements of luciferase gene induction, by using luminescence detection. Luciferase signal reflects the activation by sensitizers.

- 96 well plates are seeded. For each trial, three black well plates are seeded for the luciferase determination. One clear plate is seeded for the cytotoxicity determination. (parallel cytotoxicity measurements are conducted to assess whether gene induction levels occur at subcytotoxic concentrations). One well remains blank in each plate with no cells and no treatment to assess the background values).
- After seeding, cells are grown for 24 h in the 96 wells plates. Then medium is then removed and replaced with fresh culture medium and the test chemicals and control substances.
- Test chemicals are serially diluted to make a range of doses. Each plate contains dilutions of the test articles and dilutions of solvent/control and one well remains blank.
- Treated plates are incubated for 48 h at 37°C in the presence of 5% CO_2 in the case of KeratinoSens.
- After the exposure time, cells in the triplicate plates for luminescence readings are rinsed with phosphate buffered saline. A relevant lysis buffer added to each well for 20 min at room temperature.
- Luciferase measurements: luciferase is added to the cell lysates and plates placed in a luminometer, or the luminometer is programmed to add the luciferase substrate to each well and measure its activity.
- Cytotoxicity measurements: after 48 h the medium in the clear plate is replaced with fresh medium containing MTT (3-(4,5-dimethylthiazol-2-yl)-2,5-diphenyltetrazolium bromide) for 4 h at 37°C in the presence of 5% CO_2, then MTT medium is removed. Cell viability is measured by enzymatic conversion of the vital dye MTT [3-(4,5-dimethylthiazol-2-yl)-2,5-diphenyltetrazolium bromide] into a blue formazan salt that is quantitatively measured using a spectrometer.

- Results are calculated and reported in automated excel s
  providing information on:
  - Maximal gene induction (Imax): value observed at any co
    tion of the tested chemical and positive control is reporte
  - EC1.5 value: the concentration for which a gene in
    above the1.5 threshold (50% enhanced gene activity).
  - Cell viability values: cell viability reduced by 50% an
    and IC30 concentration values).
- Decision making: potential skin sensitizers in the KeratinoSens test
  method are:
  - Chemicals with Imax significantly higher than 1.5-fold as com-
    pared to the basal luciferase activity and the EC1.5 value is below
    1000 μM in at least two out of the three repetitions.
  - At the lowest concentration with an EC value above 1.5, the
    cellular viability should be above 70%. Otherwise chemical is
    considered cytotoxic.
  - Compounds that only induce the gene activity at cytotoxic levels
    are considered nonsensitizing skin irritants.

Data obtained from KeratinoSens in relation to data from the
LLNA has an accuracy of 77% (sensitivity 79% and specificity 72%)
for a set of 145 chemicals.[19] More detail is available on EURL
ECVAM recommendations on the KeratinoSens assay for skin sensitization testing and OECD draft proposal on this assay.[17,18]

EURL ECVAM is currently evaluating other mechanistically relevant test methods for reliability (transferability, within and between
laboratory reproducibility) and their predictive capabilities for further
use as part of integrated approaches for skin sensitization hazard
assessment.

The h-CLAT is currently under validation by ECVAM. This assay
is an in vitro skin sensitization test, based on the augmentation of
CD86 and CD54 expression in THP-1 cells after exposure to a potential sensitizer. Assay quantifies the induction of these markers, which
are associated with dendritic cells maturation in vivo, on the surface of
dendritic-cell-like cell lines.[20]

Predictive In Vivo Testing

Traditionally Guinea pig has been the most commonly used test
animal for sensitization studies. Buehler's[21] occluded path test

(without adjuvant) and Magnusson and Kligman's[22] guinea pig maximization test (using adjuvant) have been used for years, both tests measure sensitization as well as elicitation reactions. Multiple variations have been developed,[23] but the basic principles are similar and include topical application and/or intradermal injection of test material to groups of animals. Followed by a 1–2 weeks' rest period and subsequent reexposure in attempt to elicit the cutaneous hypersensitivity reactions. In some tests, an adjuvant is also administered to enhance (maximize) immune responses provoked by the test material.

Guinea Pig Maximization Test Method

A minimum of 10 animals used in the treatment group and at least 5 animals in the control group are used (or 20 test and 10 control animals). Induction is done initially on skin of shoulder region with three intradermal injections on day 0:

- 1/1 mixture of Freund's complete adjuvant (FCA)/water or saline
- Test substance (appropriate concentrations may be determined from a pilot study using two or three animals)
- Test substance in a 1/1 mixture of FCA/water or saline.

Control animals also receive three intradermal injections, but only vehicle is used instead of the test substance. 5–7 days later skin is painted with 0.5 mL of 10% sodium lauryl sulfate in vaseline, to create a local irritation, 24 h later test substance is applied under occlusion for 48 h.

Challenge is done on days 20–22 with reapplication of patches for 24 h and results are assessed at 48 and 72 h after challenge. Magnusson and Kligman grading scale is used for evaluation (0 = no visible change, 1 = discrete or patchy erythema, 2 = moderate and confluent erythema, 3 = intense erythema and swelling).

Buehler Test Method

A minimum of 20 animals in the treatment group and 10 animals in the control groups is used. Test material is applied on skin of flank under occlusion for 6 h vehicle is applied to the control group hair on the application sites are closely clipped. Same process is repeated total of three times within 2 weeks. Followed by 2 weeks rest and then challenge. To challenge patches are again applied for 6 h under occlusion and then evaluated at 24 and 56 h after application. If necessary rechallenge is done by repeating the same procedure 1 week later.

Little advice is available on test substance concentration and vehicle selection on OECD guideline 406.[24] Due to multiple limitations of these assays including subjective assessment of the frequency of responses, it is not possible to assess sensitizing potencies of a chemical using this assay.[3]

Local Lymph Node Assays

The LLNA developed in 1992 is based on an alternative strategy.[25] LLNA studies the induction phase of skin sensitization; skin sensitizers are identified by their ability to provoke proliferative responses in mouse local lymph node. Assay provides quantitative data suitable for dose—response assessment and requires less number of animals compared to guinea pig tests. The original LLNA test guideline (OECD; TG 429) was adopted in 2002 and updated in 2010.[26] The assay is based on the fact that sensitizers induce a primary proliferation of lymphocytes in the lymph nodes draining the site of application. This proliferation is proportional to the dose applied and provides objective data on of sensitization potentials. Radioactive labeling with (^{3}H) thymidine is done to measure cell proliferation. A minimum of four animals is used per dose group, with a minimum of three concentrations of the test substance, plus a negative control group treated with the vehicle only, as well as a positive control, if appropriate. Later modifications of LLNA introduced two nonradioactive modifications, LLNA: DA (TG 442 A)[27] and LLNA: BrdU-ELISA (TG 442 B)[28] were validated. A reduced LLNA (rLLNA) approach has been accepted, which could use up to 40% fewer animals. It can identify sensitizers but should not be used for hazard classification.

Protocol is as follows:

- A minimum of four animals is used per dose group (groups of four CBA/Ca female mice; 7–12 weeks of age) with a minimum of three concentrations of the test substance, plus a concurrent negative control group treated only with the vehicle for the test substance, and a positive control (concurrent or recent).
- Animals are treated topically on the dorsum of both ears with 25 μL of test material, or with an equal volume of the vehicle; once a day for 3 days.
- Five days after the initial exposure, all mice will be injected with 20 μCi (7.4 × 105 Bq) of tritiated (3H)-methyl thymidine via the tail vein.

- Five hours later, all animals are humanely killed; draining auricular lymph nodes are excised and processed.
- Cellular proliferation is determined by measuring incorporated radioactivity.
- When using nonradioactive modification to the LLNA:
  - LLNA: DA; Cellular proliferation is determined by measurement of the adenosine triphosphate (ATP) content by bioluminescence as an indicator of this proliferation.[27]
  - LLNA: BrdU-ELISA; Cellular proliferation is determined by measuring 5-bromo-2-deoxyuridine (BrdU) content, an analogue of thymidine. The BrdU content in DNA of lymphocytes is measured by ELISA.
- Results are calculated based on stimulation index (SI). Result is regarded positive when $SI \geq 3$. However, the strength of the dose–response, the statistical significance, and the consistency of the solvent/vehicle and PC responses may also be used; details are available on OECD; TG 429.[26]

LLNA has generated data on sensitizing activities, including potency of hundreds of chemicals.[29,30] These databases form an essential resource for further development of in vitro predictive methods.[31]

REFERENCES

1. *Commission of the European Communities.* Timetables for the phasing-out of animal testing in the framework of the 7th amendment to the cosmetics directive (council directive 76/768/EEC). <http://ec.europa.eu/consumers/sectors/cosmetics/files/doc/antest/sec_2004_1210_en.pdf> Accessed 01.06.04.

2. Basketter D, Alepee N, Casati S, et al. Skin sensitization—moving forward with non-animal testing strategies for regulatory purposes in the EU. *Regul Toxicol Pharmacol.* 2013;67(3): 531–535.

3. Basketter DA. Skin immunology and sensitisation. In: Chilcott RP, Price S, eds. *Principles and Practice of Skin Toxicology.* Singapore: John Wiley & Sons; 2008:151–168.

4. Saint-Mezard P, Krasteva M, Chavagnac C, et al. Afferent and efferent phases of allergic contact dermatitis (ACD) can be induced after a single skin contact with haptens: evidence using a mouse model of primary ACD. *J Invest Dermatol.* 2003;120(4):641–647.

5. Bos JD, Meinardi MM. The 500 Dalton rule for the skin penetration of chemical compounds and drugs. *Exp Dermatol.* 2000;9(3):165–169.

6. Gerberick F, Aleksic M, Basketter D, et al. Chemical reactivity measurement and the predicitve identification of skin sensitisers. The report and recommendations of ECVAM workshop 64. *Altern Lab Anim.* 2008;36(2):215–242.

7. Mutschler J, Gimenez-Arnau E, Foertsch L, Gerberick GF, Lepoittevin JP. Mechanistic assessment of peptide reactivity assay to predict skin allergens with kathon CG isothiazolinones. *Toxicol In Vitro.* 2009;23(3):439–446.

8. Rustemeyer T, van Hoogstraten IM, von Blomberg ME, Scheper RJ. Mechanisms of allergic contact dermatitis. In: Rustemeyer T, Elsner P, John SM, Maibach HI, eds. *Kanerva's Occupational Dermatology*. 2nd ed. 2012:113−146.

9. Iversen L, Johansen C. Inflammasomes and inflammatory caspases in skin inflammation. *Expert Rev Mol Diagn*. 2008;8(6):697−705.

10. Hoyer KK, Dooms H, Barron L, Abbas AK. Interleukin-2 in the development and control of inflammatory disease. *Immunol Rev*. 2008;226:19−28.

11. Adler S, Basketter D, Creton S, et al. Alternative (non-animal) methods for cosmetics testing: current status and future prospects—2010. *Arch Toxicol*. 2011;85(5):367−485.

12. Casati S, Worth A, Amcoff P, Whelan M. EURL ECVAM strategy for replacement of animal testing for skin sensitisation hazard identification and classification. *Joint Research Centre—Institute for Health and Consumer Protection* [EUR 25816]. 2013;2014(3/12).

13. *EURL ECVAM Nov 2013*. Direct peptide reactivity assay (DPRA) for skin sensitisation testing. <http://ihcp.jrc.ec.europa.eu/our_labs/eurl-ecvam/eurl-ecvam-recommendations/files-dpra/EURL_ECVAM_Recommendation_DPRA_2013.pdf> Accessed 04.10.14.

14. Gerberick GF, Vassallo JD, Bailey RE, Chaney JG, Morrall SW, Lepoittevin JP. Development of a peptide reactivity assay for screening contact allergens. *Toxicol Sci*. 2004;81(2):332−343.

15. Gerberick GF, Vassallo JD, Foertsch LM, Price BB, Chaney JG, Lepoittevin JP. Quantification of chemical peptide reactivity for screening contact allergens: a classification tree model approach. *Toxicol Sci*. 2007;97(2):417−427.

16. *OECD*. In vitro Skin sensitization: direct peptide reactivity assay (DPRA). <http://www.oecd.org/env/ehs/testing/OECD_draft%20TG_DPRA_13Nov2013.pdf> Accessed 04.12.14.

17. *EURL ECVAM Feb 2014*. Recommendations on KeratinoSens™ assay for skin sensitisation testing. <http://ihcp.jrc.ec.europa.eu/our_labs/eurl-ecvam/eurl-ecvam-recommendations/file-kerati/JRC_SPR_Keratinosens_Rec_17_02_2014.pdf> Accessed 04.10.14.

18. *OECD*. In vitro Skin sensitization: keratinocyte-based ARE-Nrf2 luciferase reporter gene test method—draft proposal for new test guidelines. <http://www.oecd.org/env/ehs/testing/OECD_DraftTG_Keratinosens_11Nov2013%20(2).pdf> Accessed 04.12.14.

19. Natsch A, Ryan CA, Foertsch L, et al. A dataset on 145 chemicals tested in alternative assays for skin sensitization undergoing prevalidation. *J Appl Toxicol*. 2013;.

20. Nukada Y, Ashikaga T, Miyazawa M, et al. Prediction of skin sensitization potency of chemicals by human cell line activation test (h-CLAT) and an attempt at classifying skin sensitization potency. *Toxicol In Vitro*. 2012;26(7):1150−1160.

21. Buehler EV. Delayed contact hypersensitivity in the guinea pig. *Arch Dermatol*. 1965;91:171−177.

22. Magnusson B, Kligman AM. The identification of contact allergens by animal assay. The guinea pig maximization test. *J Invest Dermatol*. 1969;52(3):268−276.

23. Andersen KE, Maibach HI. Guinea pig sensitization assays. an overview. *Curr Probl Dermatol*. 1985;14:263−290.

24. *OECD 406-1992*. OECD guidelines for testing of chemicals—skin sensitization. <http://www.oecd-ilibrary.org/docserver/download/9740601e.pdf?expires=1399005722&id=id&accname=guest&checksum=027E7128FE5A93D1A13C55AEE77CA663> Accessed 04.10.14.

25. Kimber I, Basketter DA. The murine local lymph node assay: a commentary on collaborative studies and new directions. *Food Chem Toxicol*. 1992;30(2):165−169.

26. *OECD Test No. 429—2010*. Skin sensitization: local lymph node assay. <http://www.oecd-ilibrary.org/docserver/download/9742901e.pdf?expires=1399005906&id=id&accname=guest&checksum=BB7913F89FDD9557CF6B48F7CC797929> Accessed 04.10.14.

27. *OECD Test No. 442A—2010*. Skin sensitization: local lymph node assay: DA. <http://www.oecd-ilibrary.org/environment/test-no-442a-skin-sensitization_9789264090972-en;jsessionid=2fn99d9jfq3qk.x-oecd-live-01> Accessed 04.08.14.

28. *OECD Test No. 442B—2010*. Skin sensitization: local lymph node assay: BrdU-ELISA. <http://www.oecd-ilibrary.org/docserver/download/9744211e.pdf?expires=1399006436&id=id&accname=guest&checksum=0FDBA3F39C7D0628DCA57BFCA8763844> Accessed 04.08.14.

29. Gerberick GF, Ryan CA, Kern PS, et al. Compilation of historical local lymph node data for evaluation of skin sensitization alternative methods. *Dermatitis*. 2005;16(4):157−202.

30. Kern PS, Gerberick GF, Ryan CA, Kimber I, Aptula A, Basketter DA. Local lymph node data for the evaluation of skin sensitization alternatives: a second compilation. *Dermatitis*. 2010;21(1):8−32.

31. Basketter DA, Gerberick GF, Kimber I. The local lymph node assay in 2014. *Dermatitis*. 2014;25(2):49−50.

CHAPTER 5

Photoallergic Contact Dermatitis: Clinical Aspects

Sally Helen Ibbotson

DEFINITION AND BACKGROUND

Topical photocontact reactions are subdivided into phototoxic and photoallergic responses. The former are much more common, nonimmunological, and thus theoretically will occur in anyone exposed to sufficient amounts of phototoxic chemical in contact with skin and light of the appropriate wavelength. In contrast, photoallergic contact reactions are much less common and fall within the class of type IV cell-mediated hypersensitivity processes in which a photoactivated chemical or a photoproduct[1] is considered to act as either a hapten or a complete antigen and ultraviolet exposure is required for both induction and elicitation of the immune response. Prior sensitization to the drug or chemical is required for photoallergic contact dermatitis to occur and thus this will not occur on first exposure to the agent and will only occur in individuals who have previously been sensitized. Distinction between phototoxicity and photoallergy is complicated as some drugs and chemicals, such as chlorpromazine, may cause photosensitization via both the systemic and topical route and can cause phototoxic and photoallergic reactions. However, this chapter concentrates on topical photoallergic contact reactions and phototoxicity and systemic photosensitization will be considered elsewhere.

HISTORY

Photoallergic dermatitis was first reported in the early 1960s in association with the use of halogenated salicylanilides used as antibacterials in soaps.[2] Index cases were reported with contact or photocontact allergy to tetrachlorosalicylanilide (TCSA) and, following investigation, this led to discontinuation of use of this photoallergic chemical. However,

Applied Dermatotoxicology. DOI: http://dx.doi.org/10.1016/B978-0-12-420130-9.00005-0

subsequently many cases of photocontact allergic dermatitis to other halogenated salicylanilides and related antimicrobials were reported.[3,4] Interestingly, both contact and photocontact allergic reactions were observed with the halogenated salicylanilides and dermatitis resolved once allergen contact had ceased in most patients, although some cases of persistent photosensitivity (persistent light reactors) were reported.[2,5] Persistent light reactivity has also been reported with other agents including musk ambrette, hexachlorophene, and olaquindox[4,6–9] although, in fact, it is generally now considered that most cases of persistent light reactivity fall within the spectrum of chronic actinic dermatitis.[10] Overall there is little evidence that topical photoallergens play any role in initiating persistent photosensitivity without ongoing contact to photoallergen or indeed play a role in the majority of cases of generalized photosensitivity.

Following the epidemic of photoallergic dermatitis to halogenated salicylanilides and other antimicrobials, subsequent cases were reported in association with the use of musk ambrette which is a potent topical photoallergen and is now no longer utilized for topical cosmetic use.[11–14] Other perfumed products such as 6-methylcoumarin have been reported as topical photoallergens[15–17] and although again now no longer available for use in topical cosmetics in European countries may still be available elsewhere on a worldwide basis and thus should still be considered as potential culprits in unexplained cases. Other photoallergens, such as the phenothiazines, were also commonly implicated.[12,18–20]

Sunscreen chemicals subsequently emerged as the major group of photocontact allergens and these have their own history with *para*-aminobenzoic acid (PABA) and PABA esters initially being the most common cause of sunscreen photocontact allergy, in part due to the photoallergic potential of these chemicals but in addition due to the high tonnage use in sunscreens at that time.[15–17,21,22] Subsequent phasing out of PABA and its esters and increased use of benzophenones, cinnamates, and dibenzoylmethanes, the latter particularly for improved longer wavelength UVA photoprotection, resulted in cases of photocontact allergic dermatitis being reported to these sunscreen groups, again reflecting increasing tonnage use of products containing these chemicals.[20,23–40] In more recent years, refinements in sunscreens and increasing complexity of sunscreen chemicals in photoprotective products has resulted in the emergence of several new photocontact allergens including drometrizole, octocrylene, and the tinosorbs.[41]

While of limited use in the United Kingdom, topical nonsteroidal antiinflammatory drug use is fairly widespread in continental Europe and elsewhere, and many topical nonsteroidals, in particular ketoprofen, are strong contact and photocontact sensitizers. There may also be cross-reactivity and for example ketoprofen cross-reacts with octocrylene and the benzophenones.[41,42] Indeed, vigilance must be kept regarding the potential impact of possible cross-reactions.[43] Thus reports of NSAID contact and photocontact allergy are fairly common and at present worldwide, sunscreen chemicals, and the topical nonsteroidals are the commonest culprits of topical photoallergic reactions.[26,44]

EPIDEMIOLOGY

It is not known how frequently photocontact allergy occurs in the general population as although there are many reports of photoallergic dermatitis in selected groups referred for investigation of either photosensitivity or topical allergic reactions, studies in the general population have not been undertaken. It is well known that in subjects of normal photosensitivity using sunscreens for prevention of sunburn and the chronic adverse effects of sun exposure, that irritancy to sunscreens is much more common than photoallergy and indeed photoallergic contact dermatitis remains uncommon.[29,45,46] Wide variation has been reported in the rates of positive photocontact allergy in selected subjects referred to either contact dermatitis or photodiagnostic units for investigation; some of this variation can be attributed to referral patterns and differences in the investigation parameters used and interpretation of outcomes. In a multicenter UK study of photocontact allergy, approximately 4% of patients were found to have topical photocontact allergy,[47] although more recently in a European study, this figure had risen to 19.4%, with topical nonsteroidals and sunscreen chemicals being the main culprits.[41] This apparent increase may well be due to us considering the possibility of topical photoallergic dermatitis as a differential diagnosis more often and therefore increasingly referring for investigation or may well reflect an increase in incidence and it is not possible to know at this stage.

MECHANISMS AND PATHOGENESIS

Covered in toxicology section.

CLINICAL PRESENTATION OF PHOTOALLERGIC CONTACT DERMATITIS

Patients will usually present with a history of, or indeed active, dermatitis on photoexposed sites and detailed history taking regarding systemic and topical drug and chemical use is essential. Often the main differential diagnoses to consider are other causes of photosensitivity including chronic actinic dermatitis and polymorphic light eruption. Airborne contact allergic dermatitis is another relatively common differential to consider. It is important also to keep the possibility of topical photoallergic dermatitis in mind in patients with a preexisting diagnosis of a photosensitivity disorder as these are patients who more regularly use sunscreens and certainly some are predisposed to the development of contact and photocontact allergic reactions, such as those with chronic actinic dermatitis. Thus, if a patient with a known photosensitivity disorder experiences deterioration of their symptoms and signs without clear explanation, then investigation for the development of topical allergic or photoallergic dermatitis should be considered.[27,28,32,48–50]

Morphology of the skin reactions in topical photoallergy is usually of a dermatitis and it is important to recognize that while the reactions will usually occur on photoexposed sites, there may be spread to covered site involvement and the distinction with photoexposed site cutoff may not be very clear.

INVESTIGATION

Photopatch testing is the investigation of choice to diagnose topical photoallergic dermatitis, and the methods and indications for photopatch testing were reviewed by the British Photodermatology Group in a workshop setting in 1995.[20] Photopatch testing is in general not useful for investigating suspected phototoxic skin reactions and indeed is of limited use for the investigation of systemic drug allergy as the interpretation and relevance of the results can be unclear.

The clinical indications for photopatch testing are patients who present with an exposed site dermatitis during the summer months or indeed any exposed site dermatitis or those with a history of a sunscreen reaction or a history of a reaction to a topical nonsteroidal antiinflammatory drug or other photoallergic chemical and of course to consider photopatch testing in patients with deterioration in a preexisting photosensitivity disease.[44]

PHOTOPATCH TESTING TECHNIQUE

Irradiation Source and Dosimetry

The wavelength dependence for an induction of photocontact allergy is known for only a few allergens,[51–56] although several are no longer clinically relevant. It appears that for photocontact allergy to at least some of the sunscreens and nonsteroidal antiinflammatory drugs[57,58] that UVA wavelengths are those that are mainly implicated in photoallergy. Indeed there is no evidence to support an added role for the use of UVB during photopatch testing.[59] Importantly in practical terms, the use of UVA radiation for photopatch testing is most appropriate as shorter wavelengths of UVB will likely elicit erythema at doses below those required to trigger photoallergy. Interestingly, for sunscreens, the chemical must be within the skin when activated by UVA to cause the photocontact reaction and preirradiation prior to skin application will not elicit the response.[60]

It is important that the UVA irradiation source has a well characterized emission spectrum, which is stable and with uniform irradiation of the field size required for use in photopatch testing. The irradiance must be sufficiently high to allow for relatively short exposure times. In practice, most centers will use either fluorescent UVA lamps (as conventionally used for PUVA) or metal halide sources, although in photodiagnostic centers a xenon-arc source filtered and coupled to an irradiation monochromator can also be used but only for small areas. The latter would really only be used in the situation of investigating a patient with photosensitivity and where monochromator phototesting facilities are available. Fluorescent UVA sources can either be those used for whole body irradiation in PUVA or smaller units which are mainly used for hand and foot irradiation. It is important that irradiance of the irradiation source is measured using a calibrated UVA meter to allow for accurate dosimetry.

A variety of UVA doses has been reported for use in the photopatch test technique, mainly between 5 and 10 J/cm^2. Higher rates of phototoxic positive reactions are reported in those studies where

10 J/cm^2 have been used and the clinical relevance of these is low, whereas phototoxicity is less likely to be induced at a dose of 5 J/cm^2 which is the conventional dose currently used in photopatch testing.[17,20,61,62] Reducing the dose below this may lead to false negative results.[63] If a patient is photosensitive, such as in chronic actinic dermatitis, then a UVA MED dose series should be undertaken to establish the UVA MED prior to irradiation with that source and a 50% UVA MED dose, or the dose below that required to elicit a minimal erythema, should be used in the photopatch test technique. However, phototesting in this patient group with preexisting photosensitivity disease may be difficult to undertake and if there is severe UVA sensitivity, then UVA irradiation will not be able to be used and only contact allergic reactions could be assessed. Interpretation and relevance of positive reactions in this patient group can also be problematic.

TEST PROCEDURE

Allergens are prepared and applied in the same way as conventional patch test techniques are undertaken (Figure 5.1) except that a duplicate series of allergens are applied (Figure 5.2) and one set will be

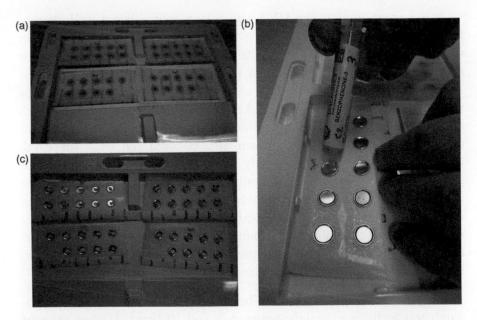

Figure 5.1 Allergen preparation: (a) Layout and numbering of aluminum Finn chambers. (b) Application of allergen to Finn chambers. (c) Allergens laid out prior to application.

irradiated and the other serves as nonirradiated control. Allergens are applied to the back avoiding the paravertebral area (Figure 5.2) and there are variables in the protocols used dependent usually on whether the investigation site is a photodiagnostic unit or a patch test clinic.[20] The main differences are that usually if patients are investigated in a patch test clinic, the allergens will be applied for 48 h before irradiation, whereas most photodiagnostic units will undertake irradiation of one set of allergens at 24 h after application as the 24 h UVA MED will be read at the same time, as this is usually undertaken at the time of allergen application (Figure 5.3). There is some retrospective evidence that a 48 h occlusion period may be more sensitive at picking up additional photoallergic reactions.[64] The allergens are applied in the usual manner on aluminum Finn chambers (Figure 5.1) and after a 24 or 48 h period, both sets of allergens are removed (Figure 5.4) and one set is randomly ascribed to be irradiated and the other is light protected (Figure 5.5). Irradiation is undertaken usually at a distance of 15 cm from the back at a dose of 5 J/cm^2.[20] In one short report, sunscreen photoallergy could be elicited for up to 96 h after application if irradiated at that time point.[65]

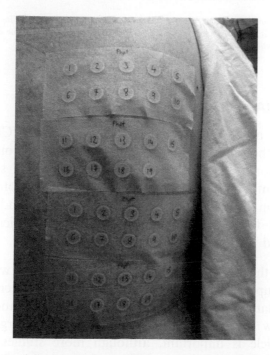

Figure 5.2 Duplicate series of allergens applied and occluded on the back avoiding the paravertebral area.

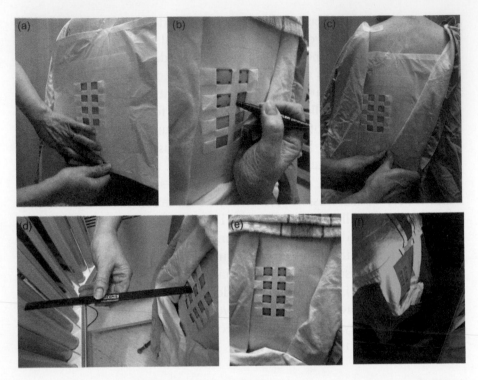

Figure 5.3 Preparation and performance of UVA MED test: (a) Application of template. (b) Marking out of test sites. (c) Gowning up of patient prior to irradiation. (d) Ensuring fixed distance for irradiation. (e) MED test sites prior to irradiation. (f) Broadband UVA irradiation to establish the MED.

TEST MATERIAL

The European multi-center photopatch test study (EMCPPTS) used a standardized photopatch test series (Table 5.1)[41] and on the basis of this, the taskforce recommended a baseline series of sunscreen and nonsteroidal antiinflammatory agents to be used in photopatch testing with the option of additional agents to be applied and the patient's own products to be applied as is if these are clinically indicated (Tables 5.2 and 5.3).[66]

It is likely that the standard photopatch test series will need to be reviewed and updated on a regular basis as new sunscreens and nonsteroidal antiinflammatories and other agents evolve and increase in usage and new allergens are identified. The photoallergen concentration and vehicle is important, although more studies are required

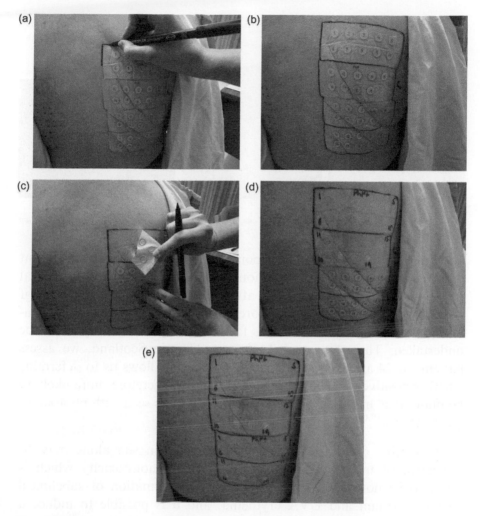

Figure 5.4 Marking out and removal of allergens: (a and b) Marking out of allergen sites prior to removal. (c and d) Removal of duplicate series of allergens. (e) Allergen application sites prior to irradiation.

to clarify the optimal conditions to induce photoallergy but avoid phototoxicity, which will likely vary dependent on the chemicals tested.[41,63,67,68]

READING AND INTERPRETATION OF RESULTS

Readings are generally undertaken immediately after irradiation and 24, 48, 72, and sometimes 96 h later, although most centers will only

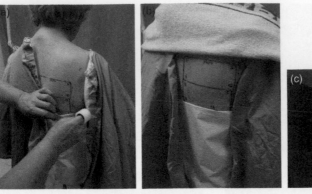

Figure 5.5 Irradiation of allergens: (a) Covering of one set of allergens. (b) Preparation of set of allergens for irradiation. (c) Irradiation with standard broadband UVA dose (usually 5 J/cm²).

undertake some of these time points and the key assessment point appears to be 48 h after irradiation. There seems to be no additional benefit of later readings 7 days after irradiation.[59] It is important if possible to obtain readings at more than one time point after irradiation as interpretation of results is facilitated if pattern analysis can be undertaken. Thus, for example, in our Unit in Scotland, we assess patients at 24 and 48 h after irradiation as this allows us to determine whether positive reactions are decrescendo and therefore more likely to be phototoxic in nature or crescendo as would be seen with photoallergic reactions.[19,69]

A positive reaction at the irradiated allergen site alone may be indicative of true photoallergy (Figure 5.6), phototoxicity which is likely to be not of clinical relevance, or a summation of subclinical chemical irritant and UVA erythema, and it is possible to induce a false positive photopatch test reaction with these parameters.[70] If reactions are strongly positive, then it is usually straightforward to distinguish between true allergy and phototoxicity or irritancy, but interpretation may be more difficult with weaker reactions and this is where having more than one time point reading to allow for pattern analysis can be helpful. Equivalent reactions at both irradiated and unirradiated or preirradiated sites usually indicates contact allergy (Figure 5.7). A positive reaction at a nonirradiated site with enhanced reaction at the irradiated site usually indicates photoaugmentation of contact allergy, although if there is a much stronger reaction at the photocontact site there may be a combination of contact and

Table 5.1 The European Multi-Center Photopatch Test Study (EMCPPTS) Photopatch Test Series[41]	
Test Agent[a]	Concentration (%)
Butyl methoxydibenzoylmethane	10
Homosalate	10
4-Methylbenzylidene camphor	10
Benzophenone-3	10
Ethylhexyl methoxycinnamate	10
Phenylbenzimidazole sulfonic acid	10
Benzophenone-4	2
Drometrizole trisiloxane	10
Octocrylene	10
Ethylhexyl salicylate	10
Ethylhexyl triazone	10
Isoamyl-p-methoxycinnamate	10
Terephthalylidene dicamphor sulfonic acid	10
Bis-ethylhexyloxyphenol methoxyphenyl triazine	10
Methylene bis-benzotriazolyl tetramethylbutylphenol	10
Diethylamino hydroxybenzoyl hexyl benzoate	10
Disodium phenyl dibenzimidazole tetrasulfonate	10
Diethylhexyl butamido triazone	10
Polysilicone-15	10
Ketoprofen	1
Etofenamate	2
Piroxicam	1
Diclofenac	5
Ibuprofen	5
Control (Petrolatum)	Not applicable

[a]*International Nomenclature of Cosmetic Ingredients (INCI) name for UV absorbers.*

photocontact allergy (Figure 5.8). A reduced reaction of a positive response at the irradiated site may indicate a degree of photoinhibition of contact allergy (Figure 5.9). If there is erythema alone at all irradiated sites, then it is difficult to interpret photopatch testing because of UVA sensitivity. If results are all negative (Figure 5.10) but there is a high index of clinical suspicion, then a technical error or suboptimal investigation parameters must be considered. Readings are scored using the same International Contact Dermatitis Research Group (ICDRG) grading system as for patch testing and the COADEX

Table 5.2 Agents Recommended for the European Photopatch Test Baseline Series[66]

Name of Agent (INCI Name for UV Absorbers)	Concentration and Vehicle
Butyl methoxydibenzoylmethane	10% pet.
Benzophenone-3	10% pet.
Benzophenone-4	2% pet.
Octocrylene	10% pet.
4-Methylbenzylidene camphor	10% pet.
Ethylhexyl methoxycinnamate	10% pet.
Isoamyl-p-methoxycinnamate	10% pet.
PABA	10% pet.
Methylene bis-benzotriazolyl tetramethylbutylphenol	10% pet.
Bis-ethylhexyloxyphenol methoxyphenyl triazine	10% pet.
Drometrizole trisiloxane	10% pet.
Terephthalylidene dicamphor sulfonic acid	10% aqua
Diethylamino hydroxybenzoyl hexyl benzoate	10% pet.
Ethylhexyl triazone	10% pet.
Diethylhexyl butamido triazone	10% pet.
Ketoprofen	1% pet.
Etofenamate	2% pet.
Piroxicam	1% pet.
Benzydamine	2% pet.
Promethazine	0.1% pet.

Table 5.3 Agents Which May be Included in Extended Series Photopatch Testing[66]

Name of Agent (INCI Name for UV Absorbers)	Concentration and Vehicle
Benzophenone-10	10% pet.
Phenylbenzimidazole sulfonic acid	10% pet.
Homosalate	10% pet.
Ethylhexyl salicylate	10% pet.
Polysilicone-15	10% pet.
Disodium phenyl dibenzimidazole tetrasulfonate	10% pet.
Dexketoprofen	1% pet.
Piketoprofen	1% pet.
Ibuprofen	5% pet.
Diclofenac	5% pet.
Fenofibrate	10% pet.
Chlorpromazine	0.1% pet.
Olaquindox	1% pet.
Triclosan	2% pet.
Trichlorocarbanilide	1% pet.

(a) 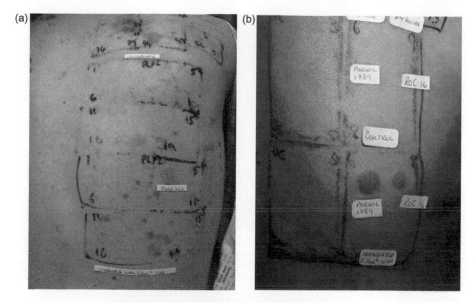 (b)

Figure 5.6 Positive photopatch test indicative of photocontact allergy: (a) Note the positive irradiated allergen site to a sunscreen chemical on the lower panel and the negative corresponding unirradiated site on the upper panel. Note also positive patch testing on the upper back which was undertaken at the same time as photopatch testing. (b) Positive photopatch test sites to sunscreen chemical and proprietary agent on lower irradiated panel and negative on unirradiated panel.

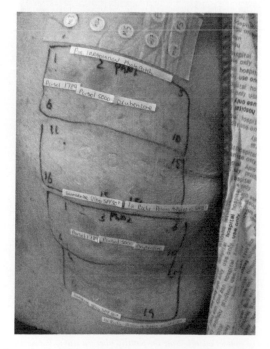

Figure 5.7 Sunscreen chemical contact allergy. Note the five positive reactions on both upper and lower allergen panels prior to irradiation and this is most suggestive of contact allergy.

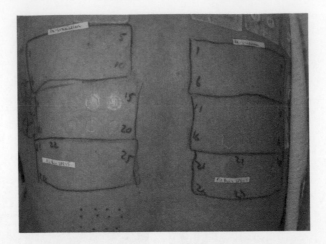

Figure 5.8 Photoaugmentation of contact allergy. Note that both the unirradiated and irradiated allergen sites for this proprietary sunscreen are positive. but that the irradiated site on the right is a stronger reaction and this is most likely indicative of photoaugmentation of contact allergy, although contact allergy combined with photocontact allergy cannot be excluded.

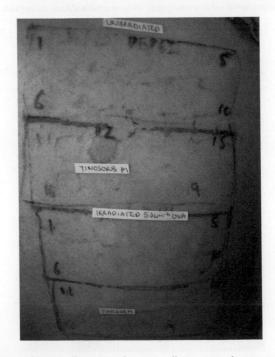

Figure 5.9 Photoinhibition of contact allergy. Note the positive allergen site to the sunscreen chemical on the unirradiated upper allergen panel, although on the corresponding irradiated lower allergen panel the same site is negative, indicating photoinhibition of contact allergy.

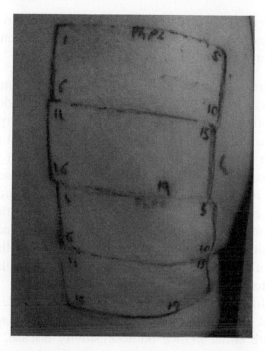

Figure 5.10 Negative photopatch testing. This may represent a true negative response or if there is a high clinical index of suspicion then technical error or suboptimal investigation parameters must be considered.

(C = current, O = old/past, A = active sensitization, D = unknown,
E = history of exposure, X = cross-reaction)[71] system can be used for
interpreting the clinical relevance of the results, which is of fundamental importance.

CURRENT COMMON TOPICAL PHOTOALLERGENS

Organic UV filters are increasingly being developed and used in sunscreen products, topical cosmetics, such as moisturizers, and in products to prevent their own photodegradation, notably shower gels and
shampoos for example. Historically, there has been a lag-time between
the introduction of new agents into the market and the emergence of
them as photocontact and contact allergens which likely reflects firstly
the increasing tonnage use of these agents over time but also the period
required for sensitization with exposure. It is important that the correct
international nomenclature of cosmetic ingredients (INCI) terminology
is given to these chemicals and that if patients are found to be allergic
or photoallergic to them that they should be given the detailed

chemical name and the INCI coding in order for them to be able to identify the prevalence in products that they may potentially use and to facilitate allergen avoidance.[38]

In 2002, a group of interested European contact dermatologists and photobiologists met to produce a consensus methodology for the photopatch testing technique, test materials, and interpretation as it was widely considered at that time that photopatch testing was underused in Europe and probably also worldwide.[72] Part of the reason for this may be because it falls between the areas of expertise of the photo-dermatologists and the contact dermatologists, but in addition at that time there was considerable variation in the technique, the components of the technique and its interpretation and there was no standardization. Detailed literature review at that time highlighted that the organic sunscreens were the most likely culprits and it was highlighted that a European photopatch test study should be undertaken.[72] Subsequently using standardized photopatch test methodology, a UK multicenter photopatch test study was undertaken in 1155 patients referred to specialist units with either known photosensitivity disease, a history of a sunscreen reaction, photoexposed site dermatitis in the summer, or a photoexposed site problem, and of these 66 (5.7%) were shown to have photoallergic contact dermatitis to at least one agent with the largest number of reactions being to benzophenone-3, butylmethoxydibenzoyl-methane, isoamyl-*p*-methoxycinnamate, and a proprietary sunscreen used as is, in addition to patients' own products (Figure 5.11). The first positive reaction to octyl triazone was also reported during the course

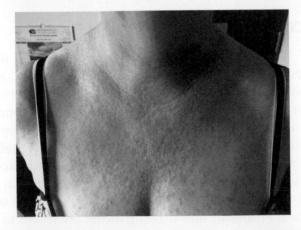

Figure 5.11 Photoexposed site dermatitis in a sunscreen user. This is an indication for photopatch testing.

of this study.[47,73] Most of the reactions in this study were considered to be of clinical relevance and most patients were unaware of the relationship to the sunscreen chemical that they had been exposed to. Additionally, the study findings indicated that there was some added value of undertaking readings 96 h after irradiation in addition to 48 h in order to increase detection of additional relevant responses.

Interestingly, in a retrospective review of patients who had been photopatch tested over a 13-year period in New York ($n = 76$), 23.2% positive photopatch tests to sunscreens but in addition antimicrobials (23.2%), medications (20.3%), fragrances (13%), plants and plant derivatives (11.6%), and pesticides (8.7%) were also shown to be still culprits.[74] However, the standard UVA dose used was 10 J/cm^2 which may have resulted in increased numbers of phototoxic reactions and 11% of the patients studied had a preexisting photosensitivity disorder. All patients were selected in this review on the basis of having suspected clinically relevant reactions but in fact several of the chemicals detected, such as the plant and plant derivatives, are phototoxic but not photoallergic, so the true relevance of some of these reactions to photoallergy is unknown.[74]

Vigilance regarding the persistence of old photoallergens[75] and the emergence of new photoallergens is important. Octocrylene is a good example of a recently recognized photoallergen and also highlights that cross-reactivity with other allergens and photoallergens should be considered. In one report of 50 cases of photoallergic contact dermatitis, octocrylene was highlighted to be a problem particularly in children and when occurring in adults was often associated with a history of photoallergy to ketoprofen and indeed cross-reactions between the benzophenones, ketoprofen, octocrylene, and also fragrances may occur.[76]

In 2007 an EMCPPTS taskforce met with the objective of standardizing photopatch testing methodology and this led to a study to determine the frequency of photoallergic contact dermatitis to organic UV absorbers and topical nonsteroidals which were in common usage in Europe.[41,72] Patients presenting with suspected photoallergic contact dermatitis were recruited and investigated using a standardized photopatch test technique to 19 UV absorbers and 5 nonsteroidal antiinflammatory agents. Patients were included if they had a photoexposed site dermatitis or a history of a sunscreen or nonsteroidal antiinflammatory drug reaction. Photopatch testing was conducted according to the

European consensus methodology[41] with allergens applied for either 24 or 48 h depending on the setup at each center. Irradiation was undertaken with 5 J/cm^2 and readings carried out immediately, 24, 48, and 72 h after irradiation or later, with the 48 h reading as the key time point. All the UV absorbers were tested at a concentration of 10% except for benzophenone-4 which was tested at 2% due to its irritancy potential.[67] The concentrations of the nonsteroidal antiinflammatory drugs were selected after a consensus discussion with experts involved at the taskforce meeting. All agents were prepared in petrolatum, with the exception of terephthalylidene dicamphor sulfonic acid which was prepared in water due its low pH, and readings were undertaken according to the ICDRG system, with relevance coded according to COADEX. Patients were also tested with up to three of their own agents, applied as is, and the study was undertaken in 30 centers prospectively over 12 European countries, recruiting 1031 patients.

Interestingly, 346 photoallergic contact reactions were reported in 200 (19.4%) subjects, which is higher than the incidence reported in the UK study.[41,47] The majority of these reactions were due to topical nonsteroidals and in particular ketoprofen ($n = 128$ subjects) and etofenamate ($n = 59$). Of the organic UV absorbers, octocrylene ($n = 41$), benzophenone-3 ($n = 37$) and butylmethoxydibenzoylmethane ($n = 18$) most frequently elicited photoallergic contact dermatitis and a cross-reaction between ketoprofen, octocrylene, and benzophenone-3 was also observed in several subjects. Allergic contact dermatitis was less commonly seen with 55 reactions in 47 (5% of subjects) with the commonest being methylene bis-benzotriazolyl tetramethylbutylphenol (Tinosorb M$^®$) ($n = 11$) and etofenamate ($n = 10$) and irritancy, photoaugmentation, and photoinhibition of allergic contact dermatitis occurred only infrequently.

The photopatch test taskforce group recommended a baseline series for photopatch testing in Europe which was based on the results of the European photopatch test study. This series consists of 20 substances consisting of absorbent sunscreens, topical nonsteroidal antiinflammatory drugs, and the topical antihistamine promethazine (Table 5.2).[66]

Fifteen additional substances were chosen to be included in an extended photopatch test series which may be important in selected cases (Table 5.3),[66] and 26 other agents were reviewed and were considered to be no longer of current relevance (Table 5.4).[66]

Table 5.4 Agents No Longer Considered Suitable for Inclusion in Photopatch Testing[66]
Name of Agent (INCI Name for UV Absorbers)
TEA salicylate
2-Ethoxyethyl-*p*-methoxycinnamate
Menthyl anthranilate
Camphor benzalkonium methosulfate
Benzylidene camphor sulfonic acid
Polyacrylamidomethyl benzylidene camphor
3-Benzylidene camphor
PEG-25 PABA
Ethylhexyl dimethyl PABA
Tiaprofenic acid
Carprofen
Suprofen
Sulfanilamide
Diphenhydramine
Chlorproethazine
Flutamide
Fluoroquinolone antibiotics
Quindoxin
Trichlorosalicylanilide
Tribromosalicylanilide
Bithionol
Fenticlor
Buclosamide
Chlorhexidine
Hexachlorophene
Sandalwood oil
6-Methylcoumarin
Musk ambrette
Wood tar
Thiourea

It was additionally considered that photopatch testing should be undertaken in parallel with patch testing to cinnamyl alcohol and decyl glucoside in order to enable relevance of positive reactions to be determined more accurately with respect to ketoprofen and methylene bis-benzotriazolyl tetramethylbutylphenol (Tinosorb M). In addition patients own products as is or in serial dilution should be considered if felt to be relevant and ongoing vigilance is required.

REGULATORY ASPECTS

Some agents are potent photoallergens and this is exemplified by keto-profen. Other chemicals are much less photoallergenic and with the introduction of new agents to market, it is important to have predictive information available on photoallergic potential.[77] Various methods have been employed in this regard including the guinea pig maximiza-tion test and other animal models.[78,79] However, there are limitations with animal models and no standardized methodology has been agreed. It is apparent that some chemicals and drugs are readily able to sensitize volunteers to photoallergic reactions and this is demon-strated with chlorproethazine[80] which was subsequently removed from the European market, and carprofen, which is restricted to veterinary use although human cases of photocontact allergy have been documen-ted.[81] If photoallergy is suspected with new agents, human volunteer testing should be considered in an attempt to avoid large-scale expo-sure of the population to potentially potent photosensitizers once these have reached the marketplace. In addition, a high index of suspicion must be maintained regarding the introduction of new chemicals with photoallergic potential and vigilance, and updating of the photopatch test batteries must be continued.

REFERENCES

1. Bruze M, Fregert S, Ljunggren B. Effect of ultraviolet irradiation of photopatch test sub-stances in vitro. *Photodermatology*. 1985;2(1):32−37.

2. Wilkinson DS. Photodermatitis due to tetrachlorsalicylanilide. *Br J Dermatol*. 1961;73:213−219.

3. Boulitronmorvan C, et al. Photoallergy to hexamidine. *Photodermatol Photoimmunol Photomed*. 1993;9(4):154−155.

4. Kumar A, Freeman S. Photoallergic contact dermatitis in a pig farmer caused by olaquindox. *Contact Dermatitis*. 1996;35(4):249−250.

5. Wilkinson DS. Patch test reactions to certain halogenated salicylanilides. *Br J Dermatol*. 1962;74:302−306.

6. Ramsay C. Transient and persistent photosensitivity due to musk ambrette. Clinical and photobiological studies. *Br J Dermatol*. 1984;111(4):423−429.

7. Kalb RE. Persistent light reaction to hexachlorophene. *J Am Acad Dermatol*. 1991;24 (2):333−334.

8. Schauder S, Schroder W, Geier J. Olaquindox-induced airborne photoallergic contact dermatitis followed by transient or persistent light reactions in 15 pig breeders. *Contact Dermatitis.* 1996;35(6):344–354.

9. Thune P, Eeglarsen T. Contact and photocontact allergy in persistent light reactivity. *Contact Dermatitis.* 1984;11(2):98–107.

10. Norris PG, Hawk JLM. Chronic actinic dermatitis—a unifying concept. *Arch Dermatol.* 1990;126(3):376–378.

11. Leow YH, et al. 2 years experience of photopatch testing in Singapore. *Contact Dermatitis.* 1994;31(3):181–182.

12. Menz J, Muller SA, Connolly SM. Photopatch testing—a six-year experience. *J Am Acad Dermatol.* 1988;18(5):1044–1047.

13. Thune P, et al. The Scandinavian multicenter photopatch study 1980–1985—final report. *Photodermatology.* 1988;5(6):261–269.

14. Wennersten G, et al. The Scandinavian multicenter photopatch study—preliminary-results. *Contact Dermatitis.* 1984;10(5):305–309.

15. DeLeo VA, Suarez SM, Maso MJ. Photoallergic contact-dermatitis—results of photopatch testing in New York, 1985 to 1990. *Arch Dermatol.* 1992;128(11):1513–1518.

16. Fotiades J, Soter NA, Lim HW. Results of evaluation of 203 patients for photosensitivity in a 7.3-year period. *J Am Acad Dermatol.* 1995;33(4):597–602.

17. Holzle E, et al. Photopatch testing: the 5-year experience of the German, Austrian, and Swiss Photopatch Test Group. *J Am Acad Dermatol.* 1991;25:59–68.

18. Guarrera M. Photopatch testing: a three-year experience. *J Am Acad Dermatol.* 1989;21(3):589–591.

19. Neumann NJ, et al. Photopatch testing: the 12-year experience of the German, Austrian, and Swiss Photopatch Test Group. *J Am Acad Dermatol.* 2000;42:183–192.

20. Ibbotson S, Farr P, Beck M. Photopatch testing-methods and indications. *Br J Dermatol.* 1997;136:371–376.

21. Dromgoole SH, Maibach HI. Sunscreening agent intolerance—contact and photocontact sensitization and contact urticaria. *J Am Acad Dermatol.* 1990;22(6):1068–1078.

22. English JSC, White IR, Cronin E. Sensitivity to sunscreens. *Contact Dermatitis.* 1987;17(3):159–162.

23. Berne B, Ros AM. 7 years experience of photopatch testing with sunscreen allergens in Sweden. *Contact Dermatitis.* 1998;38(2):61–64.

24. Davies MG, Hawk JLM, Rycroft RJG. Acute photosensitivity from the sunscreen 2-ethoxyethyl-*p*-methoxycinnamate. *Contact Dermatitis.* 1982;8(3):190–192.

25. Motley RJ, Reynolds AJ. Photocontact dermatitis due to isopropyl and butyl methoxy dibenzoylmethanes (Eusolex 8020 and Parsol 1789). *Contact Dermatitis.* 1989;21(2):109–110.

26. Bakkum RSLA, Heule F. Results of photopatch testing in Rotterdam during a 10-year period. *Br J Dermatol.* 2002;146(2):275–279.

27. Bell HK, Rhodes LE. Photopatch testing in photosensitive patients. *Br J Dermatol.* 2000;142(3):589–590.

28. Crouch RB, Foley PA, Baker CS. Letter to the Editor: The results of photopatch testing 172 patients to sunscreening agents at the photobiology clinic, St Vincent's Hospital, Melbourne. *Australas J Dermatol.* 2002;45:74.

29. Darvay A, et al. Photoallergic contact dermatitis is uncommon. *Br J Dermatol.* 2001;145:597–604.

30. De Groot A, et al. Contact allergy to 4-isopropyl-dibenzoylmethane and 3-(4'-methylbenzyli-dene) camphor in the sunscreen Eusolex 8021. *Contact Dermatitis*. 1987;16(5):249–254.

31. Goncalo M, et al. Contact and photocontact sensitivity to sunscreens. *Contact Dermatitis*. 1995;33(4):278–280.

32. Green C, Norris PG, Hawk JLM. Photoallergic contact-dermatitis from oxybenzone aggra-vating polymorphic light eruption. *Contact Dermatitis*. 1991;24(1):62–63.

33. Kimura K, Katoh T. Photoallergic contact-dermatitis from the sunscreen ethylhexyl-*p*-meth-oxycinnamate (Parsol MXC). *Contact Dermatitis*. 1995;32(5):304–305.

34. Knobler E, et al. Photoallergy to benzophenone. *Arch Dermatol*. 1989;125(6):801–804.

35. Murphy GM, White IR, Cronin E. Immediate and delayed photocontact dermatitis from iso-propyl dibenzoylmethane. *Contact Dermatitis*. 1990;22(3):129–131.

36. Journe F, et al. Sunscreen sensitization: a 5-year study. *Acta Derm Venereol*. 1999;79 (3):211–213.

37. Ricci C, Pazzaglia M, Tosti A. Photocontact dermatitis from UV filters. *Contact Dermatitis*. 1998;38:333–334.

38. Schauder S, Ippen H. Contact and photocontact sensitivity to sunscreens—review of a 15-year experience and of the literature. *Contact Dermatitis*. 1997;37(5):221–232.

39. Szczurko C, et al. Photocontact allergy to oxybenzone – 10 years of experience. *Photodermatol Photoimmunol Photomed*. 1994;10(4):144–147.

40. Thune P. Contact and photocontact allergy to sunscreens. *Photodermatology*. 1984;1:5–9.

41. Kerr AC, et al. A European multi-centre photopatch test study (EMCPPTS). *Br J Dermatol*. 2012;166(5):1002–1009.

42. Goossens A. Photoallergic contact dermatitis. *Photodermatol Photoimmunol Photomed*. 2004;20(3):121–125.

43. Pentinga SE, et al. Do "cinnamon-sensitive" patients react to cinnamate UV filters? *Contact Dermatitis*. 2009;60(4):210–213.

44. Kerr A, Ferguson J. Photoallergic contact dermatitis. *Photodermatol Photoimmunol Photomed*. 2010;26(2):56–65.

45. Foley P, et al. The frequency of reactions to sunscreens: results of a longitudinal population-based study on the regular use of sunscreens in Australia. *Br J Dermatol*. 1993;128 (5):512–518.

46. Nixon RL, Frowen KE, Lewis AE. Skin reactions to sunscreens. *Australas J Dermatol*. 1997;58(suppl):S83–S85.

47. Bryden AM, et al. Photopatch testing of 1155 patients: results of the UK multicentre photo-patch study group. *Br J Dermatol*. 2006;155(4):737–747.

48. Bilsland D, Ferguson J. Contact allergy to sunscreen chemicals in photosensitivity dermatitis actinic reticuloid syndrome (PD/AR) and polymorphic light eruption (PLE). *Contact Dermatitis*. 1993;29(2):70–73.

49. Goldermann R, et al. Contact-dermatitis from UV-A and UV-B filters in a patient with erythropoietic protoporphyria. *Contact Dermatitis*. 1993;28(5):300–301.

50. Gudmundsen KJ, et al. Polymorphic light eruption with contact and photocontact allergy. *Br J Dermatol*. 1991;124(4):379–382.

51. Cripps DJ, Enta T. Absorption and action spectra studies on bithionol and halogenated sali-cylanilide photosensitivity. *Br J Dermatol*. 1970;82(3):230–242.

52. Freeman RG, et al. Salicylanilide photosensitivity. *J Invest Dermatol*. 1970;54(2):145–149.

53. Giovinazzo VJ, et al. Photoallergic contact-dermatitis to musk ambrette—action spectra in guinea-pigs and man. *Photochem Photobiol.* 1981;33(5):773–777.

54. Freeman RG, Knox JM. The action spectrum of photocontact dermatitis. *Arch Dermatol.* 1968;97:130–136.

55. Yamada S, et al. Photoallergic contact dermatitis due to diphenhydramine hydrochloride. *Contact Dermatitis.* 1998;38(5):282.

56. Kochevar I, et al. Assay of contact photosensitivity to musk ambrette in guinea pigs. *J Invest Dermatol.* 1979;73(2):144–146.

57. Kerr A, et al. Action spectrum for etofenamate photoallergic contact dermatitis. *Contact Dermatitis.* 2011;65:115–123.

58. Collins P, Ferguson J. Photoallergic contact-dermatitis to oxybenzone. *Br J Dermatol.* 1994;131(1):124–129.

59. Pollock B, Wilkinson SM. Photopatch test method: influence of type of irradiation and value of day-7 reading. *Contact Dermatitis.* 2001;44(5):270–272.

60. Wahie S, Lloyd JJ, Farr PM. Positive photocontact responses are not elicited to sunscreen ingredients exposed to UVA prior to application onto the skin. *Contact Dermatitis.* 2007;57:273–275.

61. Duguid C, O'Sullivan D, Murphy GM. Determination of threshold UV-A elicitation dose in photopatch testing. *Contact Dermatitis.* 1993;29(4):192–194.

62. Ibbotson SH. Photo-allergy and photopatch testing. In: Ferguson J, Dover J, eds. *Photodermatology.* London;2006:72–80.

63. Hasan T, Jansen CT. Photopatch test reactivity: effect of photoallergen concentration and UVA dosaging. *Contact Dermatitis.* 1996;34(6):383–386.

64. Batchelor RJ, Wilkinson SM. Photopatch testing—a retrospective review using the 1 day and 2 day irradiation protocols. *Contact Dermatitis.* 2006;54:75–78.

65. Buckley DA, et al. Duration of response to UVA irradiation after application of a known photoallergen. *Contact Dermatitis.* 1995;33(2):138–139.

66. Gonçalo M, et al. Photopatch testing: recommendations for a European photopatch test baseline series. *Contact Dermatitis.* 2013;68(4):239–243.

67. Kerr AC, et al. A double-blind, randomized assessment of the irritant potential of sunscreen chemical dilutions used in photopatch testing. *Contact Dermatitis.* 2009;60(4):203–209.

68. Schauder S. How to avoid phototoxic reactions in photopatch testing with chlorpromazine. *Photodermatology.* 1985;2(2):95–100.

69. Neumann NJ, et al. Pattern analysis of photopatch test reactions. *Photodermatol Photoimmunol Photomed.* 1994;10:65–73.

70. Beattie PE, et al. Can a positive photopatch test be induced by subclinical irritancy or allergy plus suberythemal ultraviolet exposure? *Br J Dermatol.* 2004;51:235–240.

71. Bourke J, Coulson I, English J. Guidelines for care of contact dermatitis. *Br J Dermatol.* 2001;145(6):877–885.

72. Bruynzeel DP, et al. Photopatch testing: a consensus methodology for Europe. *J Eur Acad Dermatol Venereol.* 2004;18(6):679–682.

73. Sommer S, et al. Photoallergic contact dermatitis from the sunscreen octyl triazone. *Contact Dermatitis.* 2002;46:304–305.

74. Victor FC, Cohen DE, Soter NA. A 20-year analysis of previous and emerging allergens that elicit photoallergic contact dermatitis. *J Am Acad Dermatol.* 2010;62(4):605–610.

75. Waters AJ, et al. Photocontact allergy to PABA in sunscreens: the need for continued vigilance. *Contact Dermatitis*. 2009;60(3):173.

76. Avenel-Audran M, et al. Octocrylene, an emerging photoallergen. *Arch Dermatol*. 2010;146 (7):753−757.

77. Barratt MD, et al. Development of an expert system rulebase for the prospective identification of photoallergens. *J Photochem Photobiol B*. 2000;58(1):54−61.

78. Gerberick GF, Ryan CA. Contact photoallergy testing of sunscreens in guinea pigs. *Contact Dermatitis*. 1989;20:251−259.

79. Lovell WW, Sanders DJ. Dose-response study of ultraviolet-radiation for induction of photoallergy to tetrachlorosalicylanilide in guinea-pigs. *Photodermatol Photoimmunol Photomed*. 1990;7(5):192−197.

80. Kerr A, Woods J, Ferguson J. Photocontact allergic and phototoxic studies of chlorproethazine. *Photodermatol Photoimmunol Photomed*. 2008;24:11−15.

81. Kerr AC, et al. Occupational carprofen photoallergic contact dermatitis. *Br J Dermatol*. 2008;159(6):1303−1308.

Toxicology of Photoallergy

Golara Honari and Howard Maibach

INTRODUCTION

Photoallergy occurs when certain photoreactive allergens in skin absorb light and create an inflammatory response. These chemicals may be applied topically or diffuse into skin following systemic administration of a drug. Historically, an outbreak of photocontact dermatitis caused by an antibacterial, tetrachlorsalicylanilide, in early 1960s[1-4] raised awareness about the need to perform photosensitivity testing on drugs and skin care products. Basic principles of photoreactivity are discussed in Chapter 3. Multiple agents are known to cause photoallergic reactions including topical antimicrobials, fragrances, sunscreen ingredients such as 6-methylcoumarin, benzophenone, nonsteroidal antiinflammatory drugs (NSAIDs), promethazine, benzocaine, and p-aminobenzoic acid.

Photoallergens are haptens that require UV radiation to be able to bind carrier proteins and trigger allergic responses. While there is no validated predictive assay available to predict photoallergenicity, few in vitro and in vivo models have been developed for this purpose. Photoallergic potential of chemicals can be assessed by their potential to bind human serum albumin following UV exposure. An adjunct histidine oxidation assay can be used to differentiate photoirritants and photoallergens. This chapter reviews pathophysiology of photoallergic reactions and few available toxicology methods.

PATHOPHYSIOLOGY OF PHOTOALLERGIC REACTIONS

Photoallergic reactions are cell-mediated hypersensitivity processes requiring presensitization. Photoallergens are chemicals that absorb light and form reactive species that covalently bind proteins to transform into full allergens. A photoallergen/photohapten is basically a hapten that requires UV radiation for covalent coupling with a protein and forming a complete antigen.[5-7] This feature founded a screening in vitro assay to screen for photoallergens based on UV absorption spectrometry.[8,9] A complete photoallergen can trigger a T-cell-mediated immune response via interaction with Langerhans cells similar to a regular antigen (see Chapter 4). Photoirritants and photoallergens both form reactive species, but the chemical reactions are different, which might help further differentiate photoallergens from phototoirritants.[6,10]

Predictive *In Vitro* Testing

As mentioned earlier, a predictive method is lacking, and currently available methods are reviewed.

UV Absorption Spectrometry

Based on pathogeneses discussed above, photobinding, a well accepted property of photoallergens has founded the basis for UV absorption assays.[8,10] In this assay, test material is diluted in human serum albumin (HSA). Two set are used, one to irradiate with UV and one to keep in dark as a control. After radiation, unbound test chemical is separated from HSA by filtration. UV spectrometry is performed before and after separation. Photobinding leads to increased UV absorption. An increased absorbance more than 5% of the dark control solution is considered significant.[10]

Histidine Oxidation Assay

Histidine is a substrate for singlet oxygen.[6,11] Photooxidation of histidine has been used to screen for photoirritancy. Efficient histidine photooxidizers may be considered photoirritant rather than photoallergic. Although this isn't always the case. This assay can be used as an adjunct to UV absorption to screen for photoallergens. Phototoxic chemical, creating singlet oxygen, can lower the concentration of histidine in the test solution. Basic principles include making preparations of histidine (1 mM) and the test material (at a concentration relevant to photobinding), in sets of two. One set is exposed to UV radiation and subsequently histidine concentration is measured before and after radiation, using modified Pauly reaction.[10,12] Lovell and Jones[10] used histidine binding assay to test the 30 chemicals used in the Neutral Red Uptake Phototoxicity Test (NRU PT) validation trial; six out of seven photoallergens were identified based on the fact that they absorbed light and showed less than 5% histidine loss. Photoirritants tend to have higher ratio of histidine loss, but there are known photoirritants which do not cause histidine photooxidation.

Multiple investigators have proposed additional investigational methods and models,[7,13,14] which include assessment of photoreactivity of compounds and their ability to produce reactive oxygen species upon light exposure.[15-17] Some predictive methods include in vitro assays using human monocyte derived cells,[18] human peripheral blood monocyte derived dendritic cells (PBMDC)[19] and human skin epithelial-like cells.[20]

Predictive *In Vivo* Testing

Basic principles of in vivo studies of photoallergenicity involve two steps: sensitization (induction) and elicitation. Sensitization is induced by repeated application of the test substance followed by exposure to UV radiation (UVA and UVB). To elicit a reaction, the test compound is applied and irradiated with UVA.

Animal Models

Guinea pig is the most commonly accepted model and has been used to study photoallergy since late 1960s.[21]

Method used by Gerberick and Ryan[22] is outlined here:

- Induction: Hartley strain albino guinea pigs are used.
  - Test material is applied on depilated nuchal area. Depending on the study design test material, vehicle or empty chambers are used to apply the desired material under occlusion for about 2 h. Hill Top Chamber® have been used for occlusion. On day 1 only four intradermal injections of 1/1 mixture of Freunds Complete Adjuvant (FCA)/water outlined four corners of a 2×2 cm area for application of test substance.
  - Test materials and controls are applied three times a week for 2 weeks.
  - After each application UVA, 10 J/cm^2 is delivered. Distance from light source to skin is about 6 in. Lumbar area is covered with a foil during each irradiation.
  - Animals need to be depilated two to three times during the induction period.

- Challenge: after a 10−14-day rest period challenge is done.
  - Lumbar region is depilated and each animal will be tested either with the test material or vehicle. Two chambers containing similar ingredient are placed on the left and right side of lumbar region for 2 h.
  - After removal of patches, the test material is wiped with a moist soft cloth and only one side is exposed to UVA 10 J/cm². The other side will be protected with aluminum foil shielding.
  - Test sites are evaluated at 24 and 48 h and graded between 0 and 3. (0 = no reaction, 0.5 = slight patchy erythema, 1 = slight confluent or moderate patchy, 2 = moderate erythema, 3 = sever erythema).
  - A reaction that is more severe on the irradiated site compared to the photoprotected site indicates photocontact allergy. Similar responses on both sites indicate contact allergy.

Human Models

Photomaximization procedure[23] is used to assess topical photocontact sensitizers. This test is basically formatted after the maximization test[24] developed to identify contact sensitizers. This method also includes two steps:

- Induction: 25 consented healthy adults are enrolled in study
  - The test agents are applied under occlusion for 24 h; two times a week for 3 weeks.
  - Each application and removal is followed by UV radiation (UVA and UVB). Radiation dose is twice the individual's minimum erythema dose (MED).
- Challenge: rest period is 2 weeks then challenge is done
  - Test compounds are applied in duplicates for 24 h under occlusion.
  - After chambers are removed only one set is irradiated with UVA 4 J/cm².
  - Skin response at each site is evaluated at 24, 48, and 72 h. Results are graded from 0 to 5 (0 = no reaction, 1 = mild patchy erythema, 2 = papular reaction, moderate erythema, 3 = edema and severe erythema, 4 = edema and papules, 5 = deep erythema vesicular eruption).
  - A reaction that is more severe on the irradiated site compared to the photoprotected site indicates photocontact allergy. Similar responses on both sites indicate contact allergy.

REFERENCES

1. Wilkinson DS. Photodermatitis due to tetrachlorsalicylanilide. *Br J Dermatol.* 1961;73:213–219.

2. Wilkinson DS. Two cases of photodermatitis due to tetrachlorsalicylanilide (TCSA). *Proc R Soc Med.* 1961;54:817–818.

3. Wells GC, Harman RR. Two cases of photodermatitis due to tetrachlorsalicylanilide. *Proc R Soc Med.* 1961;54:819.

4. Calnan CD. Photodermatitis due to tetrachlorsalicylanilide. *Proc R Soc Med.* 1961;54:819–820.

5. Moser J, Hye A, Lovell WW, Earl LK, Castell JV, Miranda MA. Mechanisms of drug photobinding to proteins: photobinding of suprofen to human serum albumin. *Toxicol In Vitro.* 2001;15(4–5):333–337.

6. Lovell WW. A scheme for in vitro screening of substances for photoallergenic potential. *Toxicol In Vitro.* 1993;7(1):95–102.

7. Tokura Y. Immune responses to photohaptens: implications for the mechanisms of photosensitivity to exogenous agents. *J Dermatol Sci.* 2000;23(suppl 1):S6–S9.

8. Barratt MD, Brown KR. Photochemical binding of photoallergens to human serum albumin: a simple in vitro method for screening potential photoallergens. *Toxicol Lett.* 1985;24(1):1–6.

9. Pendlington RU, Barratt MD. Photochemical binding of photoallergens to human scrum albumin: a simple in vitro method for screening potential photoallergens. *Toxicol In Vitro.* 1990;4(4–5):307–310.

10. Lovell WW, Jones PA. Evaluation of mechanistic in vitro tests for the discrimination of photoallergic and photoirritant potential. *Altern Lab Anim.* 2000;28(5):707–724.

11 Onoue S, Hosoi K, Wakuri S, et al. Establishment and intra-/inter-laboratory validation of a standard protocol of reactive oxygen species assay for chemical photosafety evaluation. *J Appl Toxicol.* 2013;33(11):1241–1250.

12. Johnson BE, Walker EM, Hetherington AM. In vitro models for cutaneous phototoxicity. In: Marks R, Plewig G, eds. *Skin Models.* Berlin, Heidelberg: Springer-Verlag; 1986:264–281.

13. Neumann NJ, Blotz A, Wasinska-Kempka G, et al. Evaluation of phototoxic and photoallergic potentials of 13 compounds by different in vitro and in vivo methods. *J Photochem Photobiol B.* 2005;79(1):25–34.

14. Kurita M, Shimauchi T, Kobayashi M, Atarashi K, Mori K, Tokura Y. Induction of keratinocyte apoptosis by photosensitizing chemicals plus UVA. *J Dermatol Sci.* 2007;45(2):105–112.

15. Onoue S, Tsuda Y. Analytical studies on the prediction of photosensitive/phototoxic potential of pharmaceutical substances. *Pharm Res.* 2006;23(1):156–164.

16. Onoue S, Kawamura K, Igarashi N, et al. Reactive oxygen species assay-based risk assessment of drug-induced phototoxicity: classification criteria and application to drug candidates. *J Pharm Biomed Anal.* 2008;47(4–5):967–972.

17. Onoue S, Hosoi K, Wakuri S, et al. Establishment and intra-/inter-laboratory validation of a standard protocol of reactive oxygen species assay for chemical photosafety evaluation. *J Appl Toxicol.* 2012;.

18. Hoya M, Hirota M, Suzuki M, Hagino S, Itagaki H, Aiba S. Development of an in vitro photosensitization assay using human monocyte-derived cells. *Toxicol In Vitro.* 2009;23(5):911–918.

19. Karschuk N, Tepe Y, Gerlach S, et al. A novel in vitro method for the detection and characterization of photosensitizers. *PLoS One.* 2010;5(12):e15221.

20. Galbiati V, Bianchi S, Martinez V, Mitjans M, Corsini E. NCTC 2544 and IL-18 production: a tool for the in vitro identification of photoallergens. *Toxicol In Vitro*. 2014;28(1):13—17.

21. Harber LC, Targovnik SE, Baer RL. Studies on contact photosensitivity to hexachlorophene and trichlorocarbanilide in guinea pigs and man. *J Invest Dermatol*. 1968;51(5):373—377.

22. Gerberick GF, Ryan CA. Contact photoallergy testing of sunscreens in guinea pigs. *Contact Dermatitis*. 1989;20(4):251—259.

23. Kaidbey KH, Kligman AM. Photomaximization test for identifying photoallergic contact sensitizers. *Contact Dermatitis*. 1980;6(3):161—169.

24. Kligman AM. The identification of contact allergens by human assay. 3. The maximization test: a procedure for screening and rating contact sensitizers. *J Invest Dermatol*. 1966;47(5): 393—409.

CHAPTER 6

Contact Urticaria Syndrome: Clinical Aspects

Mahwish Irfan and Golara Honari

INTRODUCTION

Immediate contact reactions (ICRs) are a heterogeneous group of inflammatory responses characterized by development of inflammatory reactions within minutes of exposure to an eliciting substance.

Contact urticaria (CU) is a term introduced by Fisher in 1973, referring to a phenomenon known for years.[1] CU classically presents as a pruritic wheal or flare reaction occurring within minutes to hours of cutaneous or mucosal exposure to the offending substance. Reactions typically subside within 24 h, usually within a few hours if the offending substance is removed.[2] Contact urticaria syndrome (CUS) introduced by Maibach and Johnson in 1975 further defined reactions that happen beyond exposure sites including extracutaneous manifestations such as asthma and anaphylactoid reactions.[3] Protein contact dermatitis (PCD), first described by Hjorth and Roed-Petersen in 1976, refers to immediate eczematous eruptions on hands and forearms upon exposure to proteinaceous materials.[4]

ICRs may be caused by immunologic, nonimmunologic, or mixed or unknown mechanisms, all of which are morphologically indistinguishable. CU is not uncommon and there is an expanding list of offending agents. Reactions can be localized with erythema and burning as well as generalized involving organs other than the skin. These generalized reactions can involve the gastrointestinal tract, the respiratory tract, as well as the vascular system.[5] Immunologic CU (ICU) is usually IgE mediated, but it has also been associated with a Type 5 hypersensitivity reaction.[2,6]

Most cases of CU do not present to the dermatologist, either secondary to the obvious nature of the substance causing the problem or

Applied Dermatotoxicology. DOI: http://dx.doi.org/10.1016/B978-0-12-420130-9.00006-2

the intermittent nature of the urticaria. However, in many individuals suffering from CU, the cause may not be obvious and a high index of clinical suspicion is required for diagnosis.

CLINICAL FEATURES

CU presents with itching and/or burning, erythema, and swelling on the site of contact with the eliciting agent. Spreading of urticarial lesions, generalized involvement, or extracutaneous symptoms including angioedema and asthma is possible.

Symptoms of CUS are classified in four stages, as described by Amin and Maibach,[7] as outlined in Table 6.1. Stage 1 is localized urticaria, dermatitis, or nonspecific symptoms of itching, tingling, or burning. Stage 2 is generalized urticaria and can be seen in ICU and less commonly in nonimmunologic contact urticaria (NICU). Stage 3 involves extracutaneous symptoms including bronchial asthma, rhinitis, conjunctivitis, oropharyngeal symptoms, and gastrointestinal symptoms. Stage 4 comprises anaphylactoid reactions. Stages 3 and 4 are seen in ICU.[9]

PCD is considered by many authors as a part of CUS.[10,11] Patients typically present with chronic eczematous dermatitis, mainly located on hands and forearms, and sometimes on face. Symptoms of itching and stinging as well as erythema, swelling, vesicle formation followed by eczematous eruptions develop within minutes of handling the culprit, mainly proteinaceous products. Skin examination after acute crisis shows chronic hand dermatitis, chronic paronychia, and fingertip dermatitis.[10]

Spontaneous resolution of symptoms is generally seen when the offending agent is no longer present, but repeated exposure can lead to a chronic dermatitis. Additionally, some cases of CU have been

Table 6.1 The Contact Urticaria Syndrome (Von Krogh and Maibach)[8]	
Stage 1	Localized urticaria
	Dermatitis
	Nonspecific symptoms: itching, burning, tingling
Stage 2	Generalized urticaria
Stage 3	Extracutaneous involvement: rhino-conjunctivitis, bronchospasm, orolaryngeal symptoms (lip swelling, hoarseness, difficulty swallowing), gastrointestinal symptoms (nausea, vomiting, diarrhea, cramps)
Stage 4	Anaphylactoid reactions (shock)

associated with allergic contact dermatitis. Notably, previous dermatitis may put individuals at a higher risk for ICRs. This may be due to a compromised skin barrier that allows for penetration of allergens, specifically protein macromolecules, which are the most common causes of ICRs.[12]

MECHANISMS AND COMMON CAUSES

Nonimmunologic CU

NICU is the most common and least severe form of CU. Serious systemic reactions do not occur. Symptoms vary based on site and mode of exposure as well as the concentration and vehicle of the substance itself. NICU occurs without any previous sensitization and in almost all individuals exposed to the allergen. It is thought that the release of vasoactive mediators such as histamine and bradykinin is from direct damage to dermal vessels or a nonantibody-mediated release of these substances. An inhibitory effect of oral nonsteroidal antiinflammatory drugs (NSAIDs) such as aspirin and indomethacin and topical formulations of NSAIDs such as diclofenac gel and naproxen gel on NICU is suggestive that prostaglandins may play a role in this reaction.[6] Additionally, it has been reported that prostaglandin D2 is released during NICU reactions to benzoic acid, sorbic acid, and nicotinic acid esters.[7] Antihistamines do not inhibit reactions to substances causing NICU. The complete mechanism of NICU however is poorly understood.[5]

Allergens causing NICU are usually low-molecular-weight chemicals that easily cross the barrier of the skin. For this reason, NICU can occur on intact skin.[11] Different body sites show different susceptibility to NICU, with the face being the most sensitive followed by the antecubital fossa, upper back, upper arm, volar forearm, and lower back.[13] There is also variation between individuals in their susceptibility to develop NICU.[5]

Substances used as fragrances, preservatives, and flavoring agents are the primary offending agents that cause NICU. These can be found in foods, soft drinks, soaps, perfumes, shampoos, mouthwashes, creams, and ointments. Some of these agents include benzoic acid, dimethylsulfoxide, cobalt chloride, sodium benzoate, sorbic acid, cinnamic acid, cinnamic aldehyde, *Myroxylon pereirae*, acetic acid, butyric acid, and isopropyl alcohol.[6,14]

Immunologic CU

ICU is far less common than NICU and requires presensitization. It is a Type 1 hypersensitivity reaction requiring IgE-mediated response to a substance, most commonly a protein. These IgE antibodies react with IgE receptors on mast cells, eosinophils, basophils, and Langerhans cells among other cell types. When the allergen penetrates the skin and interacts with these IgE complexes, degranulation occurs and multiple substances including histamine, proteases, proteoglycans, and exoglycosidases are released from cells and lead to the wheal and flare response. Additionally, synthesis of leukotrienes, prostaglandins, and platelet activating factors also occurs as a result of the IgE–allergen interaction and lead to vascular permeability. If massive amounts of these substances are released, either due to widespread exposure or highly sensitized individuals, extracutaneous reactions as well as anaphylaxis can occur.[11] Reaction of IgE with Langerhans cells is thought to mediate eczematous reactions, as Langerhans cells are antigen presenting cells that present antigens to Type 2 T helper cells, which can induce a delayed type hypersensitivity reaction.[15] Individuals with atopic dermatitis are at a higher risk of developing ICU.[12,16] Compared to NICU, ICU more commonly occurs on damaged skin.[17]

Proteins as well as protein–hapten complexes are the most common culprits of ICU. Further, the allergen causing ICU may be airborne. Examples include animal and pet dander.[11]

Many substances have been reported to cause ICU. Among the most common are natural rubber latex, raw meat and fish, potatoes, antibiotics, polyethylene glycol, some metals, acrylic monomers, benzoic and salicylic acids, parabens, fruits, grains, and seafood.[18] Food handlers who are exposed to many proteins may develop an ICU or eczematous reaction on the arms and hands known as PCD, which is a Type 1 allergic reaction to high molecular weight proteins. This type of reaction is also favored in patients with atopic dermatitis and patient's with a disturbed skin barrier, allowing for penetration of the proteins.[4,19,20]

Mixed or Unknown CU

The third type of CU can exhibit both allergic and nonallergic mechanisms. It can begin as a nonallergic reaction and then an individual can become sensitized to the allergen. Many times an ICU is suspected given respiratory and gastrointestinal disturbance, but no IgE can be demonstrated in the patient's blood or tissue. Additionally, many people may react on the first exposure without presensitization.[21]

Ammonium persulfate, an oxidizing agent used in hair bleaches, is an example of this mixed mechanism. Individuals have reacted to ammonium persulfate on their first exposure and passive transfer has been found to be negative favoring NICU. However, when this is tested on controls, most remain negative, arguing against the proposed mechanism of direct action on blood vessels in NICU.[22] Additionally, individuals have presented with wide-ranging symptoms including localized and generalized urticaria, allergic contact dermatitis, irritant dermatitis, asthma, rhinitis, syncope, as well as anaphylaxis. Some of these reactions appear to be immunologically mediated, whereas others are due to nonallergic histamine release.[8,23]

DIFFERENTIAL DIAGNOSIS

When evaluating CU, it is important to exclude other entities which can have a similar presentation. Table 6.2 lists the differential diagnosis of CU.

DIAGNOSIS AND TESTING

Initial Evaluation

When evaluating a patient with suspected CU to an unidentified offending substance, a comprehensive history considering occupational and nonoccupational contacts, physical examination and allergy testing should be performed.

Table 6.2 Differential Diagnosis of Contact Urticaria
Acute urticaria
Chronic urticaria
Cholinergic urticaria
Dermographism
Pressure urticaria
Solar urticaria
Urticarial vasculitis
Acquired angioedema
Hereditary angioedema
Irritant contact dermatitis
Allergic contact dermatitis

Patients may describe a wide range of symptoms from invisible symptoms of burning, itching, and tingling to systemic complaints. Symptoms without visible clinical consequence should not be overlooked as this may be the only presenting sign. It is also very important to question patients regarding extracutaneous symptoms.[8] Additionally, the sites where lesions present may be a clue in diagnosis. The relationship between work schedule and development of reactions can also be helpful in narrowing down whether the cause is occupational or not. Clinical history should include a complete personal and family history of atopic disease. Additionally, patients should be questioned regarding drug use, stress, recent travel, hobbies, and menstrual cycle. Further, details of the patient's employment should be explored, especially for individuals in high risk occupations including health-care workers, dental workers, veterinarians, gardeners, agriculture and dairy workers, hairdressers, bakers, and food workers.[24]

Given that reactions resolve within 24 h, patient's skin may appear healthy on evaluation. Moreover, patients may not present with urticaria but may have a more chronic eczematous reaction on examination.

Laboratory Testing

Radioallergosorbent test (RAST) testing may be beneficial in finding an allergen specific IgE for an offending agent for ICU but does not help in diagnosing NICU. Serum total IgE levels may be elevated in patients but does not aid in determining whether an allergen is causing the patient's symptoms.

Allergy Testing

An algorithm for skin testing in individuals suspected to have CU has been developed by von Krogh and Maibach.[8] The steps are outlined below:

1. An open application test should first be performed on healthy skin. The test substance can be applied as is or diluted. If the patient has a history of severe reactions, serial dilutions should be performed and increasing amounts of the agent should be applied to the skin. Readings should be done at 20, 40, and 60 min.
2. If negative after the first step, an open application test should be performed on slightly affected skin or previously affected skin. Application on slightly affected skin is central in diagnosing a CU

to macromolecules or PCD. Previously affected skin may also be tested as it remains in a state of increased responsiveness for a long time after a reaction compared to healthy skin. Again, serial dilutions should be used when indicated. Readings are done at 20, 40, and 60 min.

3. If negative after the second step, an occlusive application test should be applied to healthy skin. Chambers are left in place for 15 min and then readings are done at 20, 40, and 60 min.

4. If negative after the third step, an occlusive application test should be applied to slightly affected or previously affected skin. Chambers are left in place for 15 min and then readings are done at 20, 40, and 60 min. Occlusive tests are helpful in the diagnosis of nonstandardized materials such as food products. Figure 6.1 presents multiple steps of testing with slight variations from the above.

5. In some cases, more invasive testing with prick or scratch tests is used to aid with diagnosis. Positive histamine and negative saline controls should be used. For cases of ICU, negative testing in three to five controls can further support the diagnosis.

There are many other technical factors that are essential in avoiding false reactions. The choice of vehicle for testing can be important. Agents in alcoholic solutions may cause a more rapid, short-lived reaction, whereas agents in petroleum jelly have longer lasting reactions.[2] Appropriate sites for testing include the upper back, volar aspect of the forearm, and the antecubital space. Some agents such as benzoic acid will only react on the face. Prior to testing, patients should be off all NSAIDs, steroids, and antihistamines for at least 48 h, as these agents can inhibit these types of reactions as discussed above.[25,26] Further, exposure to UV light or the phenomenon of tachyphylaxis with repeated use at the same site can also lead to false-negative results. UV irradiation can suppress these reactions for up to 3 weeks, so testing in tanned skin may not be optimal.[27,28] False positive reactions most commonly occur due to dermographism.

Passive Transfer

This procedure can prove that a CU is immunologic. In this test, 0.1 mL of fresh serum obtained from an affected individual is injected into the forearm of a human volunteer or test animal. After 24 h, 0.1 mL of the offending agent is topically applied over the injection site as well as a saline control site. A positive result is a wheal and flare

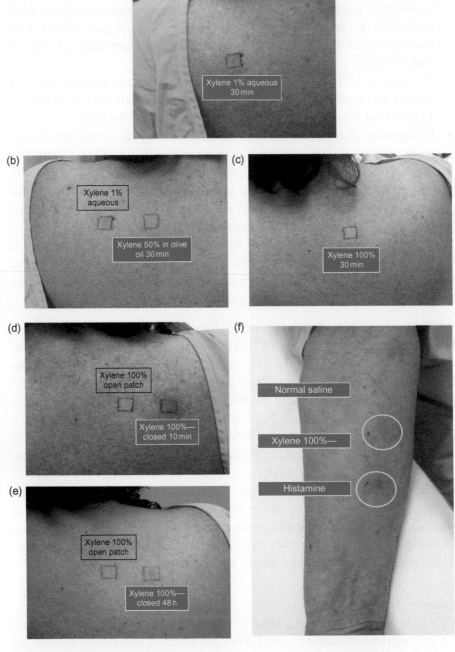

Figure 6.1 Multistep skin testing with sequential application of serial dilutions in a 58-year-old histopathology technician with sever urticarial and anaphylactoid reactions at workplace, following accidental exposure to high amounts of xylene. (a) Open application of xylene 1% aqueous solution. (b) Open application of xylene 50% in olive oil. (c) Open application of xylene 100% aqueous solution. (d) Closed patch testing with xylene 100% aqueous solution. (e) Site of closed patch testing with xylene 100% aqueous solution at 48 h. (f) Skin prick test with saline control, 100% xylene, and histamine control.

response at the donor serum injections site. This is uncommonly performed today in animals and no longer performed in humans because of infectious disease control standards.[7]

Important Considerations

Importantly, when these tests are performed, the possibility of severe extracutaneous reactions including anaphylaxis should not be forgotten. Following the above guidelines can help to reduce the occurrence of these reactions. However, life-threatening reactions during testing have been reported. For this reason, these tests should only be performed by trained personnel and when epinephrine and resuscitation equipment is available.[6]

TREATMENT AND PREVENTION

Oral antihistamines, topical corticosteroids, and a short course of oral steroids for severe symptoms can be used to help treat symptoms of CU, specifically ICU as histamine plays a fundamental role in its pathogenesis. NSAIDs do not help with ICU symptoms but can be helpful for NICU symptoms.[5,7,29] There is no pharmacological cure to reverse presensitization. Individuals with an immunological CU should carry an EpiPen as well as a medical alert tag with the name of the offending substance. Education for these individuals is of utmost importance. Signs of severe reactions should be explained thoroughly.

In occupational settings, prevention is of considerable importance. Eliminating the agent is optimal but replacing known allergens with non-cross-reactive agents can be undertaken as well. If the offending agent cannot be eliminated or replaced, proper personal protective equipment should be used and engineering controls should be implemented to avoid exposure to the allergen.[16] In severe cases where all of the above have been implemented and there is no relief, reassignment may be necessary.

DISCUSSION

It is important to properly document, diagnose, and treat CU given the negative consequences on quality of life, loss of income, work days missed, limitation of hobbies, food restrictions, and disturbance of personal relationships.[30] Additionally, given the changing work environment, novel allergens should always be considered and health-care professionals, employers, and patients need to be cognizant of this.[9]

The clinical utility of prick testing and patch testing in patients with CU related to the workplace is of substantial value. Oftentimes these patients undergo long periods of distress, decreased activities of daily life, and social isolation before their condition is properly diagnosed. Allergy testing can identify the exact substance the patient is reacting to and this can be subsequently avoided by the patient. Knowing what chemical to avoid can be very helpful in easing patient anxiety and discomfort as well as preventing severe reactions seen in ICU. Furthermore, finding the culprit substance can prevent the use of unnecessary treatments and hospitalizations for skin reactions associated with exposure if avoidance of the substance is feasible. Finally, the finding that certain chemicals cause occupational disease in many patients can lead to the institution of safer substitutes in the workplace that do not have the same adverse effects on patient health.

REFERENCES

1. Fisher AA. *Contact Dermatitis*. 2nd ed. Philadelphia, PA: Lea & Febiger; 1973.

2. Doutre M-S. Occupational contact urticaria and protein contact dermatitis. *Eur J Dermatol EJD*. 2005;15(6):419−424.

3. Maibach HI, Johnson HL. Contact urticaria syndrome. Contact urticaria to diethyltoluamide (immediate-type hypersensitivity). *Arch Dermatol*. 1975;111(6):726−730.

4. Hjorth N, Roed-Petersen J. Occupational protein contact dermatitis in food handlers. *Contact Dermatitis*. 1976;2(1):28−42.

5. Wakelin SH. Contact urticaria. *Clin Exp Dermatol*. 2001;26(2):132−136.

6. Fisher AA, Fowler JF, Rietschel RL. *Fisher's Contact Dermatitis*. Baltimore, MD: Williams & Wilkins; 1995.

7. Amin S, Lahti A, Maibach HI, eds. *Contact Urticaria Syndrome*. Boca Raton, FL: CRC Press; 1997.

8. Von Krogh G, Maibach HI. The contact urticaria syndrome—an updated review. *J Am Acad Dermatol*. 1981;5(3):328−342.

9. Wang CY, Maibach HI. Immunologic contact urticaria—the human touch. *Cutan Ocul Toxicol*. 2013;32(2):154−160. Available from: http://dx.doi.org/doi:10.3109/15569527.2012.727519.

10. Lachapelle J-M. *Patch Testing and Prick Testing: A Practical Guide*. Berlin; New York: Springer; 2003.

11. Kanerva L. *Handbook of occupational dermatology*. Available at: <http://dx.doi.org/10.1007/978-3-662-07677-4>; 2000 Accessed 22.02.14.

12. Bourrain JL. Occupational contact urticaria. *Clin Rev Allergy Immunol*. 2006;30(1):39−46. Available from: http://dx.doi.org/doi:10.1385/CRIAI:30:1:039.

13. Gollhausen R, Kligman AM. Human assay for identifying substances which induce non-allergic contact urticaria: the NICU-test. *Contact Dermatitis*. 1985;13(2):98−106.

14. Frosch PJ, Menné T, Lepoittevin J-P, eds. *Contact Dermatitis*. 4th ed Berlin; New York: Springer; 2006.

15. Najem N, Hull D. Langerhans cells in delayed skin reactions to inhalant allergens in atopic dermatitis—an electron microscopic study. *Clin Exp Dermatol*. 1989;14(3):218−222.

16. Nicholson PJ, Llewellyn D, English JS, Guidelines Development Group. Evidence-based guidelines for the prevention, identification and management of occupational contact dermatitis and urticaria. *Contact Dermatitis*. 2010;63(4):177−186. Available from: http://dx.doi.org/doi:10.1111/j.1600-0536.2010.01763.x.

17. Amaro C, Goossens A. Immunological occupational contact urticaria and contact dermatitis from proteins: a review. *Contact Dermatitis*. 2008;58(2):67−75. Available from: http://dx.doi.org/doi:10.1111/j.1600-0536.2007.01267.x.

18. McFadden J. Immunologic contact urticaria. *Immunol Allergy Clin North Am*. 2014;34(1):157−167. Available from: http://dx.doi.org/doi:10.1016/j.iac.2013.09.005.

19. Berard F, Marty J-P, Nicolas J-F. Allergen penetration through the skin. *Eur J Dermatol EJD*. 2003;13(4):324−330.

20. Smith Pease CK, White IR, Basketter DA. Skin as a route of exposure to protein allergens. *Clin Exp Dermatol*. 2002;27(4):296−300.

21. Adams RM, ed. *Occupational Skin Disease*. 3rd ed. Philadelphia, PA: Saunders; 1999.

22. Calnan CD, Shuster S. Reactions to ammonium persulfate. *Arch Dermatol*. 1963;88:812−815.

23. Fisher AA, Dooms-Goossens A. Persulfate hair bleach reactions. Cutaneous and respiratory manifestations. *Arch Dermatol*. 1976;112(10):1407−1409.

24. Guin JD. The evaluation of patients with urticaria. *Dermatol Clin*. 1985;3(1):29−49.

25. Johansson J, Lahti A. Topical non-steroidal anti-inflammatory drugs inhibit non-immunologic immediate contact reactions. *Contact Dermatitis*. 1988;19(3):161−165.

26. Lahti A, Väänänen A, Kokkonen EL, Hannuksela M. Acetylsalicylic acid inhibits non-immunologic contact urticaria. *Contact Dermatitis*. 1987;16(3):133−135.

27. Larmi E. Systemic effect of ultraviolet irradiation on non-immunologic immediate contact reactions to benzoic acid and methyl nicotinate. *Acta Derm Venereol*. 1989;69(4):296−301.

28. Larmi E, Lahti A, Hannuksela M. Ultraviolet light inhibits nonimmunologic immediate contact reactions to benzoic acid. *Arch Dermatol Res*. 1988;280(7):420−423.

29. Gimenez-Arnau A, Maurer M, De La Cuadra J, Maibach H. Immediate contact skin reactions, an update of Contact Urticaria, Contact Urticaria Syndrome and Protein Contact Dermatitis—"A Never Ending Story.". *Eur J Dermatol EJD*. 2010;20(5):552−562. Available from: http://dx.doi.org/doi:10.1684/ejd.2010.1049.

30. Kieć-Swierczyńska M, Krecisz B, Potocka A, Swierczyńska-Machura D, Dudek W, Paczyński C. Psychological factors in allergic skin diseases. *Med Pr*. 2008;59(4):279−285.

Toxicology of Contact Urticaria Syndrome

Golara Honari and Howard Maibach

INTRODUCTION

Contact urticaria syndrome as described earlier is a group of heterogeneous inflammatory reactions, happening within minutes after exposure to the offending agent. Underlying mechanisms are mainly immunologic or nonimmunologic; however, certain chemicals such as ammonium persulfate, parabens, and ethylene diamine induce contact urticaria via unknown mechanisms.[1-3]

Immunologic and nonimmunologic contact urticarias (ICU and NICU) have been studied using multiple in vivo models, but no predictive testing method has been established. The testing methods are limited to in vivo models. Majority of animal tests have investigated mechanistic aspects of nonimmunologic contact urticaria, rather than establishing predictive models. This chapter overviews test models established in assessment of nonimmunologic and immunologic contact urticaria.

PATHOGENESIS OF NONIMMUNOLOGIC CONTACT URTICARIA

NICU occurs in the absence of prior sensitization to a substance. Reactions typically remain localized to the area of contact and may present as sensory reactions such as stinging, burning, or itching to transient erythema or typical hives. The underlying mechanisms are not fully understood, but reactions are possibly caused by the release of inflammatory mediators induced by the causative agent. The causative agents either directly or indirectly via the release of inflammatory mediators can affect flow and or permeability of dermal vessels creating some of the clinical reactions. Multiple mediators and pathways are proposed and studied, including prostaglandins, histamine, mast cell activation, neurogenic inflammation, yet much remains unknown about pathogenesis.[2,4,5]

Intensity of NICU varies with the concentration of the offending agent and is also affected by properties affecting skin penetration including site of exposure.[6−9]

Benzoic acid, cinnamic acid, cinnamic aldehyde, methyl nicotinate, or dimethylsulfoxide are well-known urticants. Antihistamines do not inhibit immediate reactions to these substances, indicating histamine is not a key mediator.[10,11] However, topical application of nonsteroidal antiinflammatory gels significantly reduced erythema or edema induced by theses urticants, suggesting that NICU is partially or extensively mediated by prostaglandins.[12] Role of sensory nerves, ultraviolet light, and molecular structure of urticants have also been studied in the development of NICU.[13−16] Cinnamal is a major component of *Myroxylon pereirae* resin, both can cause nonimmunological immediate contact reactions.[17] Almost equal rate of immediate reaction to fragrance mix and *Myroxylon pereirae* in patients with either negative or positive patch test to these substances is reported, suggesting entirely separate mechanisms involved in immediate and delayed reactions.[18]

Table 6.3 lists substances associated with NICU.

ANIMAL TEST MODELS FOR NICU

Animal test methods are being used to identify urticogenic agents and at times to investigate the pathogenesis. The guinea pig ear swelling test is the best animal test available for screening NICU.[21,22] The guinea pig ear lobe resembles human skin in many respects, including the morphology and timing of the reactions. Concentrations of the eliciting substances also correlate with concentrations affecting human skin.

Protocol involves baseline measurement of ear thickness with a micrometer immediately prior to open topical application of test material followed by subsequent measurements at short periods up to 1−2 h. A positive reaction in the guinea pig ear lobe comprises erythema and edema. Quantification of the edema by measuring the change in ear thickness is an accurate, quick, and reproducible method. The maximal response is 100% increase in ear thickness and it appears 40−50 min after the application, depending on the vehicle.[16,21]

Table 6.3 Agents Producing Immediate Nonimmunologic Contact Reactions Including Contact Urticaria[9,16,19,20]

Animals

- Arthropods
- Caterpillars
- Corals
- Jellyfish
- Moths
- Sea anemones

Foods

- Cayenne pepper
- Fish
- Mustard
- Thyme

Fragrances and flavorings

- Balsam of Peru (*Myroxylon pereirae*)
- Benzaldehyde
- Cassis (cinnamon oil)
- Cinnamic acid
- Cinnamic aldehyde (cinnamal)

Medicaments and dietary supplements

- Alcohols
- Benzocaine
- Camphor
- Cantharides
- Capsaicin
- Chloroform
- Curcumin
- Dimethylsulfoxide
- Friar's balsam
- Iodine
- Methyl salicylate
- Methylene green
- Myrrh
- Nicotinic acid esters
- Resorcinol
- Tar extracts
- Tincture of benzoin
- Witch hazel

Metals

- Cobalt

Plants

- Chrysanthemum
- Christmas and Easter cacti
- Nettles
- Seaweed

Preservatives and disinfectants

- Benzoic acid
- Chlorocresol
- Formaldehyde
- Sodium benzoate
- Sorbic acid

(Continued)

Table 6.3 (Continued)
Miscellaneous
• Butyric acid
• Diethyl fumarate
• Histamine
• Hexyl nicotinate
• Pine oil
• Pyridine carboxaldehyde
• Sulfur
• Turpentine

Tachyphylaxis phenomenon is a decrease in reactivity to a nonimmunologic urticant after reapplication the following days. Length of the refractory period is about 4 days for methyl nicotinate, 8 days for diethyl fumarate and cinnamic aldehyde, and 16 days for benzoic acid, cinnamic acid, and dimethylsulfoxide.[23]

HUMAN TEST MODELS FOR NICU

The transient nature of NICU and its minor effects make human testing a reasonable option.

Open Application Test

Test substances are typically applied on skin of upper back, extensor upper arm, or forearm.

- Testing one substance at a time: 0.1 mL of the test substance is spread on a 3 × 3 cm area of the skin.
- Testing more than one substance at the same time: 0.01 mL of the test substance is applied on 1 × 1 cm area of skin.

There are marked differences between skin sites in the reactivity to the putative urticogenic agents. Petrolatum and water are the most commonly used vehicles,[24] but alcohol vehicles and added propylene glycol may enhance test sensitivity.[25,26] Test sites are evaluated at 20, 40, and 60 min in order to see the maximal response. In visual grading, scores for the erythema and edema components of the reaction (+ weak, + + moderate, + + + strong) have been used,[5,26] but objective measurements are preferred.[27] Colorimetric, spectrophotometric measurements, and laser Doppler flowmetry can be used to assess erythema and blood flow. The test is usually performed on healthy appearing skin, but when used as a diagnostic test it may be useful to test suspected agents on previously affected skin areas or slightly irritated skin. Repeated open tests on the same test site may be needed to detect weak immediate irritant reactions.[28]

Chamber Test

In chamber test, substances are applied in small aluminum chambers (Finn chamber, Epitest, Hyrylä, Finland) and fixed to the skin with porous acrylic tape similar to methods used for routine patch testing for contact allergy. The occlusion time is 15 min and the test is read at 20, 40, and 60 min. Occlusion enhances percutaneous penetration and may increase the sensitivity of the test. An advantage of the chamber test is that a smaller skin area is needed than in the open test.[29]

Practical issues to consider in testing for NICU:

- Concentration of an agent may be difficult to define, hence, dilution series are recommended. They make it possible to determine the threshold irritant concentration for that particular patient and skin area.
- It is critical that a suitable panel of control subjects is tested to assess specificity of the diagnostic test.
- Oral and topical nonsteroidal antiinflammatories may lead to false-negative results that can last up to 3 days.[12,30,31]
- Tanned skin may also cause false-negative reactions, which can last up to 2−3 weeks.[14,32]

PATHOGENESIS OF IMMUNOLOGIC CONTACT URTICARIA

ICU is far less common than NICU and requires presensitization. It is typically a Type 1 hypersensitivity reaction requiring IgE-mediated response to a substance, most commonly a protein or a protein−hapten complex. These IgE antibodies react with IgE receptors on mast cells, eosinophils, basophils, and Langerhans cells among other cell types. Upon penetration of an allergen to skin and interaction with IgE complexes, degranulation of the involved cells occur and multiple mediators such as histamine, proteases, proteoglycans, and exoglycosidases are released from cells, leading to wheal and flare response. Additionally, synthesis of leukotrienes, prostaglandins, and platelet activating factors also occurs as a result of the IgE−allergen interaction and lead to vascular permeability. Extracutaneous and anaphylactoid reactions may happen in case of massive release of these mediators in highly sensitized individuals.[33] Long-standing antigen exposures as well as interactions of IgE with Langerhans cells can mediate eczematous reactions. Langerhans cells present antigens to Type 2 T helper cells and induce a delayed type hypersensitivity reaction.[34] Individuals with atopic dermatitis are at a higher risk of developing ICU.[35,36] Compared to NICU, ICU more commonly occurs on damaged skin.[37] More details on pathogenesis of ICU and reactions with mixed or unknown mechanisms are available earlier in this chapter.

ANIMAL TEST MODELS FOR IMMUNOLOGIC CONTACT URTICARIA

Predictive models for ICU are less developed compared to NICU. Animal models have been more focused on pathogenesis rather than prediction. The guinea pig can be sensitized to a variety of chemicals that have been reported as immunologic contact urticants and proteins. Sensitization process and techniques vary when evaluating chemicals potential in skin sensitization tests, such as the guinea pig maximization test or examining the relative ability of proteins to behave as potential respiratory allergens.[38−40] In all of these procedures, once animals have been sensitized, intradermal challenges can be used to elicit an immediate hypersensitivity response in the skin. Increased vascular permeability can be visualized in the test animals if the Evans blue dye is injected prior to challenge. The reaction can be assessed in 20 min following the intradermal injection by measuring the diameter and intensity of blueness. It is not certain whether these data would predict immunologic contact urticaria in human.

Mouse has been proposed as a possible model of chemically induced immunologic contact urticaria. On the basis of the work with trimellitic anhydride (a chemical capable of causing both immediate and delayed types of hypersensitivity), an approach that involves topical application of the test chemical to BALB/c strain mice, followed by epicutaneous challenge on the ear, after 1 week has been suggested.[41] Measuring the amount of ear swelling over a 2 h course assesses urticarial reactions.[7,42] These tests have low specificity and do not provide reliable predictive information.

HUMAN TESTING FOR IMMUNOLOGIC CONTACT URTICARIA

Human testing is only done for diagnostic purposes in clinical situations. The main methods involve skin testing with the suspected substance and serological assays for specific IgE antibodies. While simple open application of a putative allergen may be sufficient, closed patch test or skin prick test may be necessary to elicit a reaction. In cases where a chemical hapten is the putative allergen, it is necessary to conjugate it to a protein prior to testing. Commonly, the protein selected is human serum albumin (HSA). Guidelines on the preparation of suitable hapten–protein conjugates are available.[43]

In patients with clinical history of sever skin reactions testing with serial dilutions of the putative allergen should be performed. Serologic tests are helpful diagnostic tools in patients with history of anaphylactic-type reactions. IgE testing has been conducted using the radioallergosorbent test (RAST), originally described over 30 years ago, is a valuable tool in identification of many immunologic urticants.[44]

SKIN TESTS FOR IMMEDIATE HYPERSENSITIVITY REACTIONS

Skin Prick Test

The standard method to detect clinically significant, immuno-globulin E (IgE)-mediated allergies. Numerous allergens are commercially available. Test materials can also be made on individual bases if not available commercially. Testing is done typically on skin of back or arm. Drops (5–10 nL) of the test material are placed 3–5 cm apart and pricked with special prick test lancet. Histamine dihydrochloride, 10 mg/mL, and the base solution serve as positive and negative controls. After puncturing the skin, drops are removed using a soft tissue. Results are read after 15–20 min by measuring the size of the wheal produced around the piercing. The longest diameter and the diameter perpendicular to it are measured. Positive results are those that are at least 3 mm or half the size of the wheal created by histamine erythema without edema is not considered positive and is not clinically relevant. When testing with nonstandard materials, it is important to test healthy controls to differentiate irritative from immunologic reactions. Prick by prick is a modification of skin prick test (SPT), in which the tested material, usually fresh foodstuff, is pricked and skin is immediately pricked using the same lancet. Prick by prick test carries a potential risk of infection particularly if fresh meat, poultry, fish, etc. are tested.[45,46]

Intradermal Test

These tests are used in clinical assessment of patients with suspected allergies and negative SPT. Test is technically more involved compared to SPT and its use is limited. Sterile solution of diluted allergen (50–100µL) is injected intradermally, using a 25-gauge needle. Intradermal injections of histamine (50–100µL) and normal saline serve as positive and negative controls. The correct way to inject the allergens would create a small papule. Injection sites will be evaluated within 20–30 min, and while a standard reading method is lacking, a wheal and flare reaction is considered positive if the largest diameter of the wheal is at least twice the diameter of the initial papule created by injection.[47]

Scratch Test

This method can be used when standard allergens are not available. In this method, superficial scratches about 5 mm long are made on skin of back or arm, using a blood lancet or a venipuncture needle. Scratches should not lead to bleeding, and lancet is not puncturing skin, hence less potential risk of infection and/or unwanted reactions compared to SPT when testing food items such as meat, fruits, vegetables, flours, and spices. Allergens are applied for 5–10 min on the scratch sites, which are 3–5 cm apart, then wiped with a soft tissue. Fresh food is applied as is, and dry allergens are mixed with normal saline. Most allergens triggering immediate responses are water-soluble.[48] Histamine dihydrochloride, 10 mg/mL, and the base solution serve as positive and negative controls. Results are read in 15–20 min after application. Similar to SPT, only reactions with erythema and edema are considered for reading and the longest diameter perpendicular to the scratch is measured and compared to histamine reactions. Reactions equal to or greater than that from histamine are considered positive and clinically relevant. It is difficult to distinguish nonimmunologic and immunologic reactions based on this method, and results should be interpreted with caution. Scratch-chamber test is a modification of scratch test when scratched skin is covered with the allergen and an 8 or 12 mm Finn chamber. Remainder of the test method and reading results is similar to scratch test.[49]

Chamber Test

Test materials are placed into a Finn chamber and, if dry, they are moistened with normal saline and applied on intact skin of back or arm. Chambers are left in place for 15 min and then readings are done at short interval up to 60 min. A weal-and-flare reaction is regarded as positive, and erythema without edema is unlikely to be positive.

Open Application Test

This test can be used in evaluation of immunologic or NICU. Although a standardized procedure is lacking proposed procedure is discussed earlier in evaluation of NICU. This test can be used when there is high clinical suspicion and careful approach is taken to avoid systemic reactions. Rub test is a modified open application test, in which the test substance is gently rubbed on the intact healthy skin or on slightly affected skin.

Skin Application Food Test

In this method, direct penetration of the allergen into skin is avoided. A solid piece of food or 0.8 mL of a liquid food is placed on a 4 cm^2 gauze and taped to skin of back for 10–30 min. The test can also be performed using patch test Finn chambers. Reactions are scored as (+ erythema, + + erythema and edema, + + + erythema and edema within the area of chamber or beyond application site) only 3 + reactions are considered positive.[50,51]

Summary

To date, no reliable method is established to predict potential immunologic or nonimmunologic urticants. Testing methods for immediate contact reactions are mostly used in clinical settings. Further understanding of mechanisms involved in pathogenesis of contact urticaria syndrome is critical for development of reliable predictive tests.

REFERENCES

1. Harvell J, Bason M, Maibach HI. Contact urticaria and its mechanisms. *Food Chem Toxicol.* 1994;32(2):103–112.

2. Amin S, Tanglertsampan C, Maibach HI. Contact urticaria syndrome: 1997. *Am J Contact Dermat.* 1997;8(1):15–19.

3. Konstantinou GN, Grattan CE. Food contact hypersensitivity syndrome: the mucosal contact urticaria paradigm. *Clin Exp Dermatol.* 2008;33(4):383–389.

4. Von Krogh G, Maibach HI. The contact urticaria syndrome. *Semin Dermatol.* 1982;1: 59–66.

5. Amin S, Lahti A, Maibach HI. Contact urticaria and the contact urticaria syndrome (immediate contact reactions). In: Zhai H, Wilhelm KP, Maibach HI, eds. *Dermatotoxicology.* 7th ed. 2008:525–536.

6. Zhai H, Zheng Y, Fautz R, Fuchs A, Maibach HI. Reactions of non-immunologic contact urticaria on scalp, face, and back. *Skin Res Technol.* 2012;18(4):436–441.

7. Lauerma AI, Maibach HI. Animal models for immunologic contact urticaria and nonimmunologic contact urticaria. In: Maibach HI, ed. *Toxicology of Skin.* USA: Taylor & Francis; 2001:252–254.

8. Basketter D, Gerberick F, Kimber I, Willis C. *Toxicology of Contact Dermatitis.* Great Britain: John Wiley & Sons; 1999.

9. Marrakchi S, Maibach HI. Functional map and age-related differences in the human face: nonimmunologic contact urticaria induced by hexyl nicotinate. *Contact Dermatitis.* 2006;55 (1):15–19.

10. Lahti A. Terfenadine does not inhibit non-immunologic contact urticaria. *Contact Dermatitis.* 1987;16(4):220–223.

11. Lahti A, McDonald DM, Tammi R, Maibach HI. Pharmacological studies on nonimmunologic contact urticaria in guinea pigs. *Arch Dermatol Res.* 1986;279(1):44–49.

12. Lahti A, Vaananen A, Kokkonen EL, Hannuksela M. Acetylsalicylic acid inhibits non-immunologic contact urticaria. *Contact Dermatitis.* 1987;16(3):133–135.

13. Gollhausen R, Kligman AM. Human assay for identifying substances which induce non-allergic contact urticaria: the NICU-test. *Contact Dermatitis.* 1985;13(2):98–106.

14. Larmi E, Lahti A, Hannuksela M. Effect of ultraviolet B on nonimmunologic contact reactions induced by dimethyl sulphoxide, phenol and sodium lauryl sulphate. *Photodermatology.* 1989;6(6):258–262.

15. Larmi E, Lahti A, Hannuksela M. Effects of infra-red and neodymium yttrium aluminium garnet laser irradiation on non-immunologic immediate contact reactions to benzoic acid and methyl nicotinate. *Derm Beruf Umwelt.* 1989;37(6):210–214.

16. Basketter D, Lahti A. Immediate contact reactions. In: Johansen JD, Frosch PJ, Lepoittevin JP, eds. *Contact Dermatitis.* 5th ed. Berlin, Heidelberg: Springer; 2011:137–153.

17. Safford RJ, Basketter DA, Allenby CF, Goodwin BF. Immediate contact reactions to chemicals in the fragrance mix and a study of the quenching action of eugenol. *Br J Dermatol.* 1990;123(5):595–606.

18. Tanaka S, Matsumoto Y, Dlova N, et al. Immediate contact reactions to fragrance mix constituents and *Myroxylon pereirae* resin. *Contact Dermatitis.* 2004;51(1):20–21.

19. Andersen F, Bindslev-Jensen C, Stahl Skov P, Paulsen E, Andersen KE. Immediate allergic and nonallergic reactions to Christmas and Easter cacti. *Allergy.* 1999;54(5):511–516.

20. Liddle M, Hull C, Liu C, Powell D. Contact urticaria from curcumin. *Dermatitis.* 2006;17 (4):196–197.

21. Lahti A, Maibach HI. An animal model for nonimmunologic contact urticaria. *Toxicol Appl Pharmacol.* 1984;76(2):219–224.

22. Lahti A, Maibach HI. Species specificity of nonimmunologic contact urticaria: guinea pig, rat, and mouse. *J Am Acad Dermatol.* 1985;13(1):66–69.

23. Lahti A, Maibach HI. Long refractory period after one application of nonimmunologic contact urticaria agents to the guinea pig ear. *J Am Acad Dermatol.* 1985;13(4): 585–589.

24. Lahti A. Immediate contact reactions. In: Rycroft RJ, Menne T, Frosch PJ, eds. *Textbook of Contact Dermatitis.* Berlin, Heidelberg, New York: Springer; 1995:62−74.

25. Lahti A, Kopola H, Harila A, Myllyla R, Hannuksela M. Assessment of skin erythema by eye, laser Doppler flowmeter, spectroradiometer, two-channel erythema meter and minolta chroma meter. *Arch Dermatol Res.* 1993;285(5):278−282.

26. Ylipieti S, Lahti A. Effect of the vehicle on non-immunologic immediate contact reactions. *Contact Dermatitis.* 1989;21(2):105−106.

27. Lahti A, Poutiainen AM, Hannuksela M. Alcohol vehicles in tests for non-immunological immediate contact reactions. *Contact Dermatitis.* 1993;29(1):22−25.

28. Hannuksela A, Niinimaki A, Hannuksela M. Size of the test area does not affect the result of the repeated open application test. *Contact Dermatitis.* 1993;28(5):299−300.

29. Hannuksela M. Skin tests for immediate hypersensitivity. In: Rycroft RJ, Menne T, Frosch PJ, eds. *Textbook of Contact Dermatitis.* Berlin, Heidelberg, New York: Springer; 1995:287−292.

30. Johansson J, Lahti A. Topical non-steroidal anti-inflammatory drugs inhibit non-immunologic immediate contact reactions. *Contact Dermatitis.* 1988;19(3):161−165.

31. Kujala T, Lahti A. Duration of inhibition of non-immunologic immediate contact reactions by acetylsalicylic acid. *Contact Dermatitis.* 1989;21(1):60−61.

32. Larmi E. Systemic effect of ultraviolet irradiation on non-immunologic immediate contact reactions to benzoic acid and methyl nicotinate. *Acta Derm Venereol.* 1989;69(4):296−301.

33. Hannuksela M. Mechanisms in contact urticaria. *Clin Dermatol.* 1997;15(4):619−622.

34. Najem N, Hull D. Langerhans cells in delayed skin reactions to inhalant allergens in atopic dermatitis—an electron microscopic study. *Clin Exp Dermatol.* 1989;14(3):218−222.

35. Nicholson PJ, Llewellyn D, English JS, Guidelines Development Group. Evidence-based guidelines for the prevention, identification and management of occupational contact dermatitis and urticaria. *Contact Dermatitis.* 2010;63(4):177−186.

36. Bourrain JL. Occupational contact urticaria. *Clin Rev Allergy Immunol.* 2006;30(1):39−46.

37. Amaro C, Goossens A. Immunological occupational contact urticaria and contact dermatitis from proteins: a review. *Contact Dermatitis.* 2008;58(2):67−75.

38. Blaikie L, Morrow T, Wilson AP, et al. A two-centre study for the evaluation and validation of an animal model for the assessment of the potential of small molecular weight chemicals to cause respiratory allergy. *Toxicology.* 1995;96(1):37−50.

39. Sarlo K, Fletcher ER, Gaines WG, Ritz HL. Respiratory allergenicity of detergent enzymes in the guinea pig intratracheal test: association with sensitization of occupationally exposed individuals. *Fundam Appl Toxicol.* 1997;39(1):44−52.

40. Magnusson B, Kligman AM. The identification of contact allergens by animal assay. The guinea pig maximization test. *J Invest Dermatol.* 1969;52(3):268−276.

41. Dearman RJ, Mitchell JA, Basketter DA, Kimber I. Differential ability of occupational chemical contact and respiratory allergens to cause immediate and delayed dermal hypersensitivity reactions in mice. *Int Arch Allergy Immunol.* 1992;97(4):315−321.

42. Lauerma AI, Fenn B, Maibach HI. Trimellitic anhydride-sensitive mouse as an animal model for contact urticaria. *J Appl Toxicol.* 1997;17(6):357−360.

43. Bernstein DI, Zeiss CR. Guidelines for preparation and characterization of chemical−protein conjugate antigens. Report of the subcommittee on preparation and characterization of low molecular weight antigens. *J Allergy Clin Immunol.* 1989;84(5 Pt 2):820−822.

44. Wide L. A RAST neutralization test for detection of blocking antibodies in serum after hyposensitization. *Int Arch Allergy Appl Immunol.* 1976;52(1−4):219−226.

45. Rance F, Juchet A, Bremont F, Dutau G. Correlations between skin prick tests using commercial extracts and fresh foods, specific IgE, and food challenges. *Allergy*. 1997;52 (10):1031−1035.

46. Heinzerling L, Mari A, Bergmann KC, et al. The skin prick test—European standards. *Clin Transl Allergy*. 2013;3(1):3-7022-3-3

47. Bindslev-Jensen C. Skin tests for immediate hypersensitivity. In: Johansen JD, Frosch PJ, Lepoittevin JP, eds. *Contact Dermatitis*. 5th ed. Berlin, Heidelberg: Springer; 2011:511−517.

48. Abou Chakra OR, Sutra JP, Poncet P, Lacroix G, Senechal H. Key role of water-insoluble allergens of pollen cytoplasmic granules in biased allergic response in a rat model. *World Allergy Organ J*. 2011;4(1):4−12.

49. Niinimaki A. Scratch-chamber tests in food handler dermatitis. *Contact Dermatitis*. 1987;16 (1):11−20.

50. Oranje AP, Van Gysel D, Mulder PG, Dieges PH. Food-induced contact urticaria syndrome (CUS) in atopic dermatitis: reproducibility of repeated and duplicate testing with a skin provocation test, the skin application food test (SAFT). *Contact Dermatitis*. 1994;31(5):314−318.

51. Oranje AP. Skin provocation test (SAFT) based on contact urticaria: a marker of dermal food allergy. *Curr Probl Dermatol*. 1991;20:228−231.

Printed and bound by CPI Group (UK) Ltd, Croydon, CR0 4YY

03/10/2024

01040420-0015